Alfred Neighbour

**The Apiary**

Bees, Bee-Hives and Bee Culture

Alfred Neighbour

**The Apiary**
*Bees, Bee-Hives and Bee Culture*

ISBN/EAN: 9783337812843

Printed in Europe, USA, Canada, Australia, Japan

Cover: Foto ©berggeist007 / pixelio.de

More available books at **www.hansebooks.com**

# THE APIARY;

OR,

# BEES, BEE-HIVES

AND

# BEE CULTURE.

# THE APIARY;

OR,

# BEES, BEE-HIVES,

AND

# BEE CULTURE:

BEING A FAMILIAR ACCOUNT OF THE HABITS OF BEES, AND THE MOST
IMPROVED METHODS OF MANAGEMENT, WITH FULL DIRECTIONS,
ADAPTED FOR THE COTTAGER, FARMER, OR
SCIENTIFIC APIARIAN.

By ALFRED NEIGHBOUR.

"Je connais beaucoup des personnes qu' aiment les abeilles, mais ne personnes
aiment les mediocrement, ils les passionnent."—GELIEU.

LONDON:
KENT AND CO., PATERNOSTER ROW;
GEO: NEIGHBOUR AND SONS,
119, REGENT STREET, AND 127, HIGH HOLBORN;
AND ALL BOOKSELLERS.
1865.

# TABLE OF CONTENTS.

## SECTION IV.

## SECTION V.—MISCELLANEOUS INFORMATION.

## APPENDIX.—TESTIMONIALS OF THE PRESS.

## ERRATA.

At second line from bottom of page 23, for " cottage " read " cottager."

At page 30, fifth line from bottom, for " last page " read " page 25."

At page 44, sixteenth line, for " *this* plates of glass" read " *thin* plates of glass."

At page 53, fourteenth line, read " bee hive " for " bee house."

## EXPLANATIONS OMITTED.

At page 62, " Hold the glass horizontally over the flame of the candle."

At page 80, " An empty hive should be placed on the stand when the living hive is removed, for the purpose of amusing returning bees. If the hive is kept in a bee-house, the entrance should be shut down until the hive is restored, when the clustered bees may be at once admitted."

# PREFACE.

Our apology for preparing a bee book is a very simple one. We are so frequently applied to for advice on matters connected with bees and bee-hives, that it seemed likely to prove a great advantage, alike to our correspondents and ourselves, if we could point to a "handy-book" of our own, which should contain full and detailed replies sufficient to meet all ordinary enquiries. Most of the apiarian manuals possess some special excellence or other, and we have no wish to disparage any of them. Yet in all, we have found a want of explanations relating to several of the more recent improvements.

It has, more especially, been our aim to give explicit and detailed directions on most subjects connected with the hiving and removing of bees; and also, to show how, by judicious application of the "depriving" system, the productive powers of the bees may be enormously increased.

We need say little here as to the interest that attaches to the apiary as a source of perennial pleasure for the amateur naturalist. Many of the hives and methods of management are described with a direct reference to this class of bee-keepers; so that, besides plain and simple directions suitable for cottagers with their ordinary hives, this work will be found to include instructions useful for the scientific apiarian, or, at least, valuable, for those who desire to gain a much wider acquaintance with the secrets of bee-keeping than is now usually possessed. We would lay stress on the term "acquaintance," for there is nothing in the management of the various bar-and-frame hives which is at all difficult when frequent practice has rendered the bee-keeper familiar with them. Such explicit directions are herein given as to how the right operations may be performed at the right times, that a novice may at once commence to use the modern hives. The word "new-fangled" has done good service for the indolent and prejudiced; but we trust that our readers will be of a very different class. Let them give a fair trial to the modern appliances for the humane and depriving system of bee-keeping, and they will find offered to them an entirely new field of interest and observation. At present, our continental neighbours far surpass us as bee-masters; but we trust that the season of 1865, if the summer be fine, will prove a turning point in the course of English bee-keeping. There is little doubt that a greater number of intelligent and influential persons in this country will become bee-keepers than has ever been the case before.

Our task would have lost half its interest, did we not hope that it would result in something beyond the encouragement of a refined and interesting amusement for the leisurely classes. The social importance of bee keeping as a source of pecuniary profit for small farmers and agricultural labourers, has never been appreciated as it deserves. Yet these persons will not, of themselves, lay aside the bungling and wasteful plan of destroying the bees, or learn without being taught the only proper method, that of deprivation. Their educated neighbours when once interested in bee-keeping, will be the persons to introduce the more profitable system of humane bee-keeping. The clergy, especially, as permanent residents in the country, may have great influence in this respect. There is not a rural or suburban parish in the kingdom in which bee-keeping might not be largely extended, and the well being of all but the very poorest inhabitants would be greatly promoted. Not only would the general practice of bee-keeping add largely to the national resources, but that addition would chiefly fall to the share of those classes to whom it would be of most value. Moreover, in the course of thus adding to their income, the uneducated classes would become interested in an elevating and instructive pursuit.

It is curious to observe that honey, whether regarded as a manufactured article or as an agricultural product, is obtained under economical conditions of exceptional advantage. If regarded as a manufactured article, we notice that there is no outlay required for "labour," nor any expense for "raw material." The industrious labourers are eager to utilize all their strength: they never "combine" except for the benefit of their master, they never "strike" for wages, and they provide their own subsistence. All that the master manufacturer of honey has to do financially, is to make a little outlay for "fixed capital" in the needful "plant" of hives and utensils—no "floating capital" is needed. Then, on the other hand, if we regard honey as an agricultural product, it presents as such a still more striking contrast to the economists' theory of what are the "requisites of production." Not only is there no outlay needed for wages and none for raw material, but there is nothing to be paid for "use of a natural agent." Every square yard of land in the United Kingdom may come to be cultivated, as in China, but no proprietor will ever be able to claim "rent" for those "waste products" of the flowers and leaves, which none but the winged workers of the hive can ever utilize.

The recent domestication in England of the Ligurian or "Italian Alp" bee adds a new and additional source of interest to bee-culture. We have, therefore, gone pretty fully into this part of the subject; and believe that what is here published with regard to their introduction embodies the most recent and reliable information respecting them that is possessed by English apiarians.*

---

* Some of our apiarian friends may be inclined to be discouraged from cultivating the Ligurian bees in consequence of the liability to their becoming hybridised when located in proximity to the black bees. We can dispel these fears by stating that we have not unfrequently found that hybrid queens possess the surprising fecundity of the genuine Italian ones, whilst the English stocks in course of time become strengthened by the infusion of foreign blood.

We are under many obligations for the advice and assistance that we have on many occasions received from Mr. T. W. Woodbury, of Exeter, whose apiarian skill is unrivalled in this country. Our acknowledgments are also due to Mr. Henry Taylor, author of an excellent " Bee-Keeper's Manual," for his help and counsel during the earlier years of our apiarian experience. Both the before-mentioned gentlemen have freely communicated to us their contrivances and suggestions, without thought of fee or reward for them. In common with most recent writers on bee-culture, we are necessarily largely indebted to the standard works of Huber and succeeding apiarians. From the more recent volume of the Rev. L. L. Langstroth we have also obtained useful information. But having ourselves of later years had considerable experience in the manipulation and practical management of bees, we are enabled to confirm or qualify the statement of others, as well as to summarize information gleaned from many various sources. We should state that our thanks are due to Mr. W. Martin Wood, who has rendered us valuable aid in the arrangement of this work.

Let it be understood that we have no *patented devices* to push: we are free to choose out of the many apiarian contrivances that have been offered of late years, and we feel perfectly at liberty to praise or blame as our experience warrants us in doing. It does not follow that we necessarily disparage hives which are not described herein; we have sought, as much as possible, to indicate the *principles* on which *good hives* must be constructed, whatever their outward size or shape. All through the work, we have endeavoured to adopt the golden rule of "submission to nature," by reference to which all the fancied difficulties of bee-keeping may be easily overcome. In none of the attempts of men to hold sway over natural objects, is the truth of Bacon's leading doctrine more beautifully illustrated than in the power that the apiarian exercises in the little world of bees.

There are one or two literary peculiarities of this work to which we ought to refer. It will be noticed that the numbers affixed to the names of the various hives and utensils do not follow in regular order. Those numbers refer to our Illustrated Catalogue, with which most English apiarians have long been familiar, and we could not alter them without great inconvenience.

Some persons may consider we have used too many poetical quotations in a book dealing wholly with matters of fact. We trust, however, that an examination of the extracts will at once remove that feeling of objection.

We venture to hope that the following pages contain many valuable hints and interesting statements which may tend to excite increased and renewed attention to the most useful and industrious of all insects. Although bees have neither reason nor religion for their guide, yet from them man may learn many a lesson of virtue and industry, and may even draw from them thoughts suggestive of trust and faith in God.

We beg leave to conclude our preface, and introduce the subject, by the following extract from Shakespeare,—who, without doubt, kept bees in that garden at Stratford wherein he used to meditate :—

> " So work the honey bees ;
> Creatures, that by a rule in Nature, teach
> The art of order to a peopled kingdom.
> They have a king and officers of sorts ;
> Where some, like magistrates, correct at home ;
> Others, like merchants, venture trade abroad ;
> Others, like soldiers, armèd in their stings,
> Make boot upon the Summer's velvet buds,
> Which pillage they, with merry march, bring home
> To the tent royal of their emperor :
> Who, busied in his majesty, surveys
> The singing masons building roofs of gold ;
> The civil citizens kneading up the honey ;
> The poor mechanic porters crowding in
> Their heavy burdens at his narrow gate ;
> The sad-eyed justice, with his surly hum,
> Delivering o'er to executors pale
> The lazy, yawning drone."—

SHAKSPEARE'S *Henry V.*, *Act* 1., *Scene* 2.

# THE APIARY;

OR,

## BEES, BEE-HIVES, & BEE CULTURE.

———•o°⚬°o•———

THERE are two classes of persons for whom bee-culture should have a strong interest, and two distinct purposes for which the pursuit may be followed. First, there is the cottager or small farmer who, in thousands of instances, might add considerably to his income by bee-keeping; and secondly, there is the man of " retired leisure " and refinement, who, in the personal tendance of an apiary would find an easy and interesting occupation, and one which could not fail to quicken his faculties of general scientific observation. Moreover, in contemplating the wonderful skill, industry, and prevision of his insect artizans, the bee-keeper would find in his apiary constant illustrations of creative wisdom.

Amongst the humbler classes in the rural districts, the neglect of bee-keeping is to be attributed to an exaggerated idea of the trouble needful for the care of a few hives, and also to ignorance of the easier and more profitable methods of modern management. Many of the wealthier country or suburban residents, also, are averse to the personal trouble which they fancy needful in keeping an apiary; and, perhaps, some gentlemen are more afraid than they would like to own of that very efficient weapon of defence with which the honey bee is provided. But the prejudices against bees are quite unnecessary; bees are as tractable as they are intelligent, and it is the purpose of this little book to show that bee-culture is an easy and safe, as well as a deeply interesting, pursuit. Possibly, also, some who do us the favour to read our detailed explanations, will see how the rural clergyman, or the benevolent landlord, who keeps an apiary of his own, may be of signal service to his poorer neighbours in explaining to them the mysteries of bee-keeping.

B

## I. THE INHABITANTS OF THE HIVE.

Every hive or bee colony comprises three distinct classes of bees, each class having functions peculiar to itself, and which are essential to the well-being of the whole community. As each bee knows its own proper duties, they all work harmoniously and zealously together, for the common weal. Certain apparent exceptions to the good-fellowship of the bees will be hereafter noticed, but those arise out of essential conditions in the social economy of the bee community.

The three classes of bees are :—the queen bee, with the pupæ or embryos intended for queens ; the working bees ; and the drones or male bees.

THE QUEEN.—Dr. Evans* introduces the queen bee to our notice thus :—

> First of the throng, and foremost of the whole,
> One 'stands confest the sovereign and the soul.'

The queen may very readily be distinguished from the rest of the bees by the greater length of her body, and the comparative shortness of her wings ; her legs are longer, and are not furnished with either brushes or baskets as those of the working bee, for being constantly fed by the latter, she does not need those implements ; the upper surface of her body is of a brighter black than the other bees, whilst her colour underneath is a yellowish brown ; her wings, which do not extend more than half the length of her body, are sinewy and strong ; her long abdomen tapers nearly to a point ; her head is rounder, her tongue more slender, and not nearly so long, as that of the working bee, and her sting is curved ; her movements are measured and majestic, as she moves in the hive the other bees form a circle round her, none venturing to turn their backs upon her, but all anxious to show that respect and attention due to her rank and station. Whenever in the exercise of her sovereign will the queen wishes to travel amongst her subjects, she experiences no inconvenience from overcrowding ; although the part of the hive to

---

* Dr. Evans,—who may be styled the poet-laureate of the bees,—lived at Shrewsbury, where he practised as a physician. His poem on bees is written with great taste and careful elaboration, and it describes the habits of bees with a degree of accuracy only attainable after continuous scientific observation.

which she is journeying may be the most populous, way is immediately made, the common bees tumbling over each other to get out of her way, so great is their anxiety not to interfere with the royal progress.

It is the chief function of the queen to lay the eggs from which all future bees originate, the multiplication of the species being the purpose of her existence, and she follows it up with an assiduity similar to that with which the workers construct combs or collect honey. A queen is estimated to lay in the breeding season from 1,500 to 2,000 eggs a-day, and in the course of one year is supposed to produce more than 100,000 bees. This is indeed a vast number; but when there is taken into consideration the great number required for swarms, the constant lessening of their strength by death in various ways, and the many casualties attending them in their distant travels in search of the luscious store, it does not seem that the case is over stated.

In a Glass Unicomb Hive,—which we shall hereinafter describe, —all the movements of the queen-bee may be traced; she may be seen thrusting her head into a cell to discover whether it be occupied with an egg or honey, and if empty, she turns round in a dignified manner and inserts her long body—so long, that she is able to deposit the egg at the bottom of the cell; she then passes on to another, and so continues industriously multiplying her laborious subjects. It not unfrequently happens when the queen is prolific, and if it be an early season, that many eggs are wasted for want of unoccupied cells; for in that case the queen leaves them exposed at the bottom of the hive when they are greedily devoured by the bees.

The queen-bee, unlike the great majority of her subjects, is a stayer at home; generally speaking, she only quits the hive twice in her life. The first occasion is on the all-important day of her marriage, which always takes place at a great height in the air, and generally on the second or third day of her princess-life; she never afterwards leaves the hive, except to lead off an emigrating swarm. Evans, with proper loyalty, has duly furnished a glowing epithalamium for the queen-bee :—thus,

> When noon-tide Sirius glares on high,
> Young love ascends the glowing sky,
> From vein to vein swift shoots prolific fire,
> And thrills each insect fibre with desire ;
> Then Nature to fulfil thy prime decree,
> Wheels round in wanton rings, the courtier Bee ;

Now shyly distant, now with bolder air,
He woos and wins the all-complying fair;
Through fields of ether, veiled in vap'ry gloom
They seek, with amorous haste, the nuptial room;
As erst the immortal pair, on Ida's height,
Wreath'd round their noon of joy, ambrosial night.

The loyalty and attachment of bees to their queen is one of their most remarkable characteristics; they constantly supply her with food, and fawn upon and caress her, softly touching her with their antennæ, a favour which she occasionally returns. When she moves about the hive, all the bees through whom she successively passes pay her the same homage; those whom she leaves behind in her track close together, and resume their accustomed occupations.

The majestic deportment of the queen-bee and the homage paid to her is, with a little poetic licence, thus described by Evans :—

But mark of royal port and awful mien,
Where moves with measured pace the insect Queen!
Twelve chosen guards, with slow and solemn gait,
Bend at her nod, and round her person wait.

This homage is, however, only paid to matron queens. Whilst they continue princesses, they receive no distinctive marks of respect. Dr. Dunbar, the noted Scotch apiarian, observed a very striking instance of this whilst experimenting on the combative qualities of the queen-bee. "So long," says he, "as the queen which survived the rencontre with her rival, remained a virgin, not the slightest degree of respect or attention was paid her—not a single bee gave her food; she was obliged, as often as she required it, to help herself; and in crossing the honey cells for that purpose, she had to scramble, often with difficulty, over the crowd, not an individual of which got out of her way, or seemed to care whether she fed or starved : but no sooner did she become a mother, than the scene was changed, and all testified towards her that most affectionate attention, which is uniformly exhibited to fertile queens."

The queen-bee, though provided with a sting, never uses it on any account, except in combat with her sister queens. But she admits of no rival to her throne; almost her first act on coming forth from the cell, is an attempt to tear open and destroy the cells containing the pupæ of princesses likely to become competitors. Should it so happen that another queen of similar age does exist in the hive at

the same time, the two are speedily brought into contact with each
other in order to fight it out and decide by a struggle, mortal to one
of them, which is to be the ruler;—the stronger of course is
victorious, and remains supreme. This, it must be admitted, is a
wiser method of settling the affair than it would be to range the
whole band under two distinct banners, and so create a civil war,
killing and destroying each other for matters with which they indi-
vidually have little or no concern: for the bees care not which
queen it is, as long as they are certain of having one to rule over
them and perpetuate the community.

After perusing the description given above of the attachment
of bees to their queen, it may be easy to imagine the consternation
a hive is thrown into when deprived of her presence. The bees first
make a diligent search for their monarch in the hive, and then afterwards rush forth in immense numbers to seek her. When such a
commotion is observed in an apiary, the experienced bee-master will
repair the loss by giving a queen: the bees have generally their
own remedy for such a calamity, in their power of raising a new
queen from amongst their larvæ; but if neither of these means be
available, the whole colony dwindles and dies. The following is the
method by which working bees provide a successor to the throne
when deprived of their queen by accident, or in anticipation of the
first swarm, which is always led by the old queen:—

They select, when not more than three days old, an egg or grub
previously intended for a worker-bee, and then enlarge the cell so
selected by destroying the surrounding partitions; they thus form
a royal cradle, in shape very much like an acorn cup inverted. The
chosen embryo is then fed liberally with a peculiar description of
nurture, called by naturalists "royal jelly"—a pungent food, prepared
by the working bees exclusively for those of the larvæ that are
destined to become candidates for the honour of royalty. Should
a queen be forcibly separated from her subjects, she resents the
interference, refuses food, pines, and dies.

The whole natural history of the queen-bee is in itself a subject
that will well repay for continuous study. Those who desire to
follow it, we would refer to the complete works of HUBER—the
greatest of apiarians,—SWAMMERDAM, BEVAN, LANGSTROTH, &c.
The observations upon the queen-bee needful to verify the above mentioned facts can only be made in hives constructed for the purpose,

of which our "Unicomb Observatory Hive" is one of the best. In ordinary hives the queen is scarcely ever to be seen; where there are several rows of comb, she invariably keeps between them, both for warmth and to be more secure from danger. The writer has frequently observed in stocks which have unfortunately died, that the queen was one of the last to expire; and she is always more difficult to gain possession of than other bees, being by instinct taught that she is indispensable to the welfare of her subjects.

The queen enjoys a far longer life than any of her subjects, her age generally extending to four or even five years. The drones, which are mostly hatched in the early spring, seldom live more than three or four months, even if they should escape the sting of the executioner, to which they generally fall victims. The worker-bee, it is now a well-ascertained fact, lives from six to eight months, in no case exceeding the latter; so that we may reckon that the bees hatched in April and May expire about the end of the year, and it is those of the autumn who carry on the duties of the hive until the spring and summer, that being the time when the greatest number of eggs are laid. The population of a hive is very small during the winter, in comparison with the vast numbers gathering produce in the summer,—produce which they themselves live to enjoy but for a short period. So that not only, as of old, may lessons of industry be learned from bees, but they also teach self-denial to mankind, since they labour for the community rather than for themselves. Evans, in describing the age of bees, thus paraphrases the well known couplet of Homer in allusion to the fleeting generations of men :—

> Like leaves on trees, the race of bees is found,
> Now green in youth, now withering on the ground;
> Another race the spring or fall supplies,
> They droop successive, and successive rise.

THE DRONE.—The drones are male bees; they possess no sting, are more hairy and larger than the common bee, and may be easily distinguished by their heavy motion, thick-set form, and louder humming. Evans thus describes the drones :—

> Their short proboscis sips,
> No luscious nectar from the wild thyme's lips;
> From the lime leaf no amber drops they steal,
> Nor bear their grooveless thighs the foodful meal:

On others' toils in pampered leisure thrive,
The lazy fathers of the industrious hive;
Yet oft, we're told, these seeming idlers share
The pleasing duties of parental care;
With fond attention guard each genial cell,
And watch the embryo bursting from the shell.

But Dr. Evans had been "told" what was not correct when he sought to dignify drones with the office of "nursing fathers,"—that task is undertaken by the younger of the working-bees. No occupation falls to the lot of the drones in gathering honey, nor have they the means provided them by nature for assisting in the labours of the hive. The drones are the progenitors of working bees, and nothing more; so far as is known, that is the only purpose of their short existence.

In a well-populated hive the number of drones is computed at from one to two thousand. "Naturalists," says Huber, "have been extremely embarrassed to account for the number of males in most hives, and which seem only a burden to the community, since they appear to fulfil no function. But we now begin to discern the object of nature in multiplying them to such an extent. As fecundation cannot be accomplished within the hive, and as the queen is obliged to traverse the expanse of the atmosphere, it is requisite that the males should be numerous, that she may have the chance of meeting some one of them in her flight. Were only two or three in each hive, there would be little probability of their departure at the same instant with the queen, or that they would meet her in their excursions; and most of the females might thus remain sterile." It is important for the safety of the queen-bee that her stay in the air should be as brief as possible: her large size, and the slowness of her flight, render her an easy prey to birds. It is not now thought that the queen always pairs with a drone of the *same* hive, as Huber seems to have supposed. Once impregnated,—as is the case with most insects,—the queen-bee continues productive during the remainder of her existence. It has, however, been found that though old queens cease to lay worker eggs, they may continue to lay those of drones. The swarming season being over, that is about the end of July, a general massacre of the "lazy fathers" takes place. Dr. Bevan, in the "Honey Bee," observes on this point, "the work of the drones being now completed, they are regarded as useless consumers of the fruits of others' labour, love is at once converted into hate, and a general proscription takes place. The unfortunate victims evidently perceive their danger, for they are never, at this time, seen resting in one place, but darting in and out of the hive with the utmost precipitation, as if in fear of being seized."

Their destruction is thought, by some, to be caused by their being harassed until they quit the hive; but Huber says he ascertained that the death of the drones was caused by the stings of the workers. Supposing the drones come forth in May, which is the average period of their being hatched, their destruction takes place somewhere about the commencement of August, so that three months is the usual extent of their existence; but should it so happen that the usual development of the queen has been retarded, or that the hive has in any case been deprived of her, the massacre of the drones is deferred. But in any case, the natural term of the life of drone bees does not exceed four months, so that they are all dead before the winter, and are not allowed to be useless consumers of the general store.

THE WORKER BEE.—The working bees form, by far, the most numerous class of the three kinds contained in the hive, and least of all require description. They are the smallest of the bees, are dark brown in colour or nearly black, and much more active on the wing than are either drones or queens. The usual number in a healthy hive varies from twelve to thirty thousand; and, previous to swarming, exceeds the larger number. The worker-bee is of the same sex as the queen, but is only partially developed. Any egg of a worker-bee,—by the cell being enlarged, as already described, and the "royal jelly" being supplied to the larva,—may be hatched into a mature and perfect queen. This, one of the most curious facts connected with the natural history of bees, may be verified in any apiary by most interesting experiments, which may be turned to important use. With regard to the supposed distinctions between "nursing" and working bees, it is now agreed that it only consists in a division of labour,—the young workers staying at home to feed the larvæ until they are themselves vigorous enough to range the fields in quest of supplies. But, for many details of unfailing interest, we must again refer our readers to the standard works on bees that have already been named.

THE EGGS OF BEES.—It is necessary that some explanation

should be given as to the existence of the bee before it emerges from the cell.

The eggs of all the three kinds of bees when first deposited are of an oval shape, and of a bluish-white colour. In four or five days the egg changes to a worm, and in this stage is known by the names of larva or grub, in which state it remains four to six days more; during this period it is fed by the nurse-bees with a mixture of farina and honey, a constant supply of which is given to it: the next transformation is to the nymph or pupa form; the nurse-bees now seal up the cell with a preparation similar to wax; and and then the pupa spins round itself a film or cocoon, just as a silkworm does in its chrysalis state. The microscope shows that this cradle-curtain is perforated with very minute holes, through which the baby-bee is duly supplied with air. No further attention on the part of the bees is now requisite except a proper degree of heat, which they take care to keep up, a position for the breeding cells being selected in the centre of the hive where the temperature is likely to be most congenial.

Twenty-one days after the egg is first laid (unless cold weather should have retarded it) the bee quits the pupa state, and nibbling its way through the waxen covering that has enclosed it, comes forth a winged insect. In the Unicomb Observatory Hive, the young bees may distinctly be seen as they literally fight their way into the world, for the other bees do not take the slightest notice, nor afford them any assistance. We have frequently been amused in watching the eager little new-comer, now obtruding its head, and anon compelled to withdraw into the cell, to escape being trampled on by the apparently unfeeling throng, until at last it has succeeded in making its exit. The little grey creature, after brushing and shaking itself, enters upon its duties in the hive, and in a day or two may be seen gathering honey in the fields—some say on the day of its birth,—thus early illustrating that character for industry, which has been proverbial, at least, since the days of Aristotle, and which has in our day been rendered familiar even to infant minds through the nursery rhymes of Dr. Watts.

INCREASE OF BEES.—Every one is familiar with the natural process of "swarming," by which bees provide themselves with fresh space and seek to plant colonies to absorb their increase of population. But the object of the bee-master is to train and educate his bees, and in so doing he avoids much of the risk and trouble which is incurred by allowing the busy folk to follow their own devices. The various methods for this end adopted by apiarians all come under the term of the "depriving" system; and they form part of the great object of humane and economical bee-keeping, which is to save the bees alive instead of slaughtering them as under the old clumsy system. A very natural question is often asked,—how it is that upon the depriving system, where our object is to prevent swarming, the increase of numbers is not so great as upon the old plan? It will be seen that the laying of eggs is performed by the queen only, and that there is but one queen to each hive; so that where swarming is prevented, there remains only one hive or stock, as the superfluous princesses are not allowed to come to maturity. Our plan of giving additional store-room will, generally speaking, prevent swarming; this stay-at-home policy, we contend, is an advantage, for instead of the loss of time consequent upon a swarm hanging out preparatory to flight, all the bees are engaged in collecting honey, and that at a time when the weather is most favourable and the food most abundant. Upon the old system, the swarm leaves the hive simply because the dwelling has not been enlarged at the time when the bees are increasing. The emigrants are always led off by the old queen, leaving either young or embryo queens to lead off after swarms, and to furnish a mistress for the old stock, and carry on the multiplication of the species. Upon the antiquated and inhuman plan where so great a destruction takes place by the brimstone match, breeding must, of course, be allowed to go on to its full extent to make up for such sacrifices. Our chief object under the new system is to obtain honey free from all extraneous matter. Pure honey cannot be gathered from combs where storing and breeding are performed in the same compartment. For fuller explanations on this point, we refer to the various descriptions of our improved hives in a subsequent section of this work.

There can now be scarcely two opinions as to the uselessness of the rustic plan of immolating the poor bees after they have striven through the summer so to "improve each shining hour." The ancients in Greece and Italy took the surplus honey and spared the bees, and now for every intelligent bee-keeper there are ample appliances wherewith to attain the same results. Mr. Langstroth quotes from the German the following epitaph which, he says, " might be properly placed over every pit of brimstoned bees :"—

### Here Rests,

CUT OFF FROM USEFUL LABOUR,

A COLONY OF

## INDUSTRIOUS BEES,

BASELY MURDERED

BY ITS

## UNGRATEFUL AND IGNORANT
## OWNER.

And Thomson, the poet of " The Seasons," has recorded an eloquent poetic protest against the barbarous practice, for which, however, in his day there was no alternative :—

> Ah, see, where robbed and murdered in that pit,
> Lies the still heaving hive ! at evening snatched,
> Beneath the cloud of guilt-concealing night,
> And fix'd o'er sulphur ! while, not dreaming ill,
> The happy people, in their waxen cells,
> Sat tending public cares.
> Sudden, the dark, oppressive steam ascends,
> And, used to milder scents, the tender race,
> By thousands, tumble from their honied dome
> Into a gulf of blue sulphureous flame !

It will be our pleasing task in subsequent chapters to show "a more excellent way."

## SWARMING.

The spring is the best period at which to open an apiary, and swarming-time is a good starting point for the new bee-keeper. The period known as the swarming season is during the months of May and June. With a very forward stock, and in exceedingly fine weather, bees do occasionally swarm in April. The earlier the swarm the greater is its value. If bees swarm in July, they seldom gather sufficient to sustain themselves through the winter ; though, by careful feeding, they may easily be kept alive, if hived early in the month.

The cause of a swarm leaving the stock-hive is, that the population has grown too large for it. Swarming is a provision of nature for remedying the inconvenience of overcrowding, and is the method whereby the bees seek for space in which to increase their stores. By putting on " super hives," the required relief may, in many cases, be given to them ; but should the multiplication of stocks be desired, the bee-keeper will defer increasing the space until the swarm has issued forth. In May, when the spring has been fine, the queen-bee is very active in laying eggs, and the increase in a strong healthy hive is so prodigious that emigration is necessary, or the bees would cease to work.

It is now a well established fact that the old queen goes forth with the first swarm, preparation having been made to supply her place as soon as the bees determine upon the necessity of a division of their commonwealth. Thus the sovereignty of the old hive, after the first swarm has issued, devolves upon a young queen.

As soon as the swarm builds combs in its new abode, the emigrant-queen, being impregnated and her ovaries full, begins laying eggs in the cells, and thereby speedily multiplies the labourers of the new colony. Although there is now amongst apiarians no doubt that the old queen quits her home, there is no rule as to the composition of the swarm—old and young alike depart. Some show unmistakable signs of age by their ragged wings, others their extreme youth by their lighter colour; how they determine which shall stay and which shall go has not yet been ascertained. In preparation for flight, bees commence filling their honey bags, taking sufficient, it is said, for three days' sustenance. This store is needful, not only for food, but to enable the bees to commence the secretion of wax and the building of combs in their new domicile.

On the day of emigration the weather must be fine, warm, and clear, with but little wind stirring; for the old queen, like a prudent matron, will not venture out unless the day is in every way favorable. Whilst her majesty hesitates, either for the reasons we have mentioned, or because the internal arrangements are not sufficiently matured, the bees will often fly about or hang in clusters at the entrance of the hive for two or three days and nights together, all labour meanwhile being suspended. The agitation of the little folk is well described by Evans :—

> See where, with hurried step, the impassioned throng
> Pace o'er the hive, and seem, with plaintive song,
> T' invite the loitering queen ; now range the floor,
> And hang in cluster'd columns from the door ;

Or now in restless rings around they fly,
Nor spoil thy sip, nor load the hollowed thigh;
E'en the dull drone his wonted ease gives o'er,
Flaps his unwieldly wings, and longs to soar.

But when all is ready, a scene of the most violent agitation takes place ; the bees rush out in vast numbers, forming quite a dark cloud as they traverse the air.

The time selected for the departure of the emigrants is generally between 10 a.m. and 3 p.m. ; most swarms come off within an hour of noon. It is a very general remark that bees choose a Sunday for swarming, and probably this is because then greater stillness reigns around. It will not be difficult to imagine that the careful bee-keeper is anxious to keep a strict watch, lest he should lose such a treasure when once it takes wing. The exciting scene at a bee-swarming has been well described by the apiarian laureate :—

Up mounts the chief, and, to the cheated eye,
Ten thousand shuttles dart along the sky ;
As swift through æther rise the rushing swarms,
Gay dancing to the beam their sunbright forms ;
And each thin form, still lingering on the sight,
Trails as it shoots, a line of silver light.
High poised on buoyant wing, the thoughtful queen,
In gaze attentive, views the varied scene,
And soon her far-fetched ken discerns below ;
The light laburnam lift her polished brow,
Wave her green leafy ringlets o'er the glade.
Swift as the falcon's sweep, the monarch bends
Her flight abrupt ; the following host descends
Round the fine twig, like clustered grapes they close
In thickening wreaths, and court a short repose.

In many country districts it is a time-honoured custom for the good folks of the village to commence on such occasions a terrible noise of tanging and ringing with frying pan and key. This is done with the absurd notion that the bees are charmed with the clangorous din, and may by it be induced to settle as near as possible to the source of such sweet sounds. This is, however, quite a mistake ; there are other and better means for the purpose. The practice of ringing was originally adopted for a different and far more sensible object, viz., for the purpose of giving notice that a swarm had issued forth, and that the owner was anxious to claim the right of following, even though it should alight on a neighbour's premises. It would be curious to trace how this ancient ceremony has thus got corrupted from the original design.

In case the bees do not speedily after swarming manifest signs of settling, a few handfuls of sand or loose mould may be thrown up in the air so as to fall among the winged throng; they mistake this for rain, and then very quickly determine upon settling. Some persons squirt a little water from a garden engine in order to produce the same effect.

There are, indeed, many ingenious devices used by apiarians for decoying the swarms. Mr. Langstroth mentions a plan of stringing dead bees together, and tying a bunch of them on any shrub or low tree upon which it is desirable that they should alight; another plan is, to hang some black woven material near the hives, so that the swarming bees may be led to suppose they see another colony, to which they will hasten to attach themselves. Swarms have a great affinity for each other when they are adrift in the air; but, of course, when the union has been effected, the rival queens have to do battle for supremacy. A more ingenious device than any of the above, is by means of a mirror to flash a reflection of the sun's rays amongst a swarm, which bewilders the bees, and checks their flight. It is manifestly often desirable to use some of these endeavours to induce early settlement, and to prevent, if possible, the bees from clustering in high trees or under the eaves of houses, where it may be difficult to hive them.

Should prompt measures not be taken to hive the bees as soon as the cluster is well formed, there is danger of their starting on a second flight; and this is what the apiarian has so much to dread. If the bees set off a second time, it is generally for a long flight, often for miles, so that in such a case it is usually impossible to follow them, and consequently a valuable colony may be irretrievably lost.

Too much care cannot be exercised to prevent the sun's rays falling on a swarm when it has once settled. If exposed to heat in this way, bees are very likely to decamp. We have frequently stretched matting or sheeting on poles so as to intercept the glare, and thus render their temporary position cool and comfortable.

Two swarms sometimes depart at the same time and join together; in such a case, we recommend that they be treated as one by putting them into a hive as before described, taking care to give abundant room, and not to delay affording access to the super hive or glasses. They will settle their own notions of sovereignty by one queen destroying the other. There are means of separating two swarms if done at the time; but the operation is a formidable one, and does not always repay even those most accustomed to such manipulation.

With regard to preparations for taking a swarm, our advice to the bee-keeper must be the reverse of Mrs. Glass's notable injunction as to the cooking of a hare. Some time before you expect to take a swarm, be sure to have a proper hive in which to take it, and also every other requisite properly ready. Here we will explain what was said in the introduction as to the safety of moving and handling bees. A bee-veil or dress will preserve the most sensitive from the possibility of being stung. This article, which may be bought with the hives, is made of net close enough to exclude bees, but open enough for the operator's vision. It is made to go over the hat of a lady or cap of a gentleman ; it can be tied round the waist, and has sleeves fastening at the wrist. A pair of photographer's india rubber gloves completes the full dress of the apiarian, who is then invulnerable even to enraged bees. But bees when swarming are in an eminently peaceful frame of mind; having dined sumptuously, they require to be positively provoked before they will sting. Yet there may be one or two foolish bees who, having neglected to fill their honey bags, are inclined to vent their ill-humour on the kind apiarian. When all is ready, the new hive is held or placed in an inverted position under the cluster of bees, which the operator detaches from their perch with one or two quick shakes ; the floorboard is then placed on the hive, which is then slowly turned up on to its base, and it is well to leave it a short time in the same place, in order to allow of stragglers joining their companions.

If the new swarm is intended for transportation to a distance, it is as well for it to be left at the same spot until evening, provided the sun is shaded from it : but if the hive is meant to stand in or near the same garden, it is better to remove it within half an hour to its permanent position, because so eager are newly-swarmed bees for pushing forward the work of furnishing their empty house, that they sally forth at once in search of materials.

A swarm of bees in their natural state contains from 10,000 to 20,000 insects, whilst in an established hive they number 40,000 and upwards. 5,000 bees are said to weigh one pound ; a good swarm will weigh from three to five. pounds. We have known swarms not heavier than 2½ pounds, that were in very excellent condition in August as regards store for the winter.

Hitherto, all our remarks have had reference to first or " prime " swarms; these are the best, and when a swarm is purchased such should be bargained for.

Second swarms, known amongst cottage bee-keepers as " casts," usually issue from the hive nine or ten days after the first has departed. It is not always that a second swarm issues, so much depends on the strength of the stock, the weather, and other causes; but should the bees determine to throw out another, the first hatched queen in the stock-hive is prevented by her subjects from destroying the other royal princesses, as she would do if left to her own devices. The consequence is that, like some people who cannot have their own way, she is highly indignant; and when thwarted in her purpose, utters, in quick succession, shrill, angry sounds, much resembling " *peep, peep*," commonly called " piping," but which more courtly apiarians have styled the *vox regalis*.

This royal wailing continues during the evening, and is sometimes so loud as to be distinctly audible many yards from the hive. When this is the case, a swarm may be expected either on the next day, or at latest within three days. The second swarm is not quite so chary of weather as the first; it was the *old* lady who exercised so much caution, disliking to leave home except in the best of summer weather.

In some instances, owing to favourable breeding seasons and prolific queens, a third swarm issues from the hive, this is termed a " colt ;" and in remarkable instances, even a fourth, which in rustic phrase is designated a " filly." A swarm from a swarm is called a " maiden " swarm, and according to bee theory, will again have the old queen for its leader.

The bee-master should endeavour to prevent his labourers from swarming more than once; his policy is rather to encourage the industrious gathering of honey by keeping a good supply of " supers " on the hives. Sometimes, however, he may err in putting on the supers too early or unduly late, and the bees will then swarm a second time, instead of making use of the store-rooms thus provided. In such a case, the clever apiarian, having spread the swarm on the ground, will select the queen, and cause the bees to go back to the hive from whence they came. This operation requires an amount of apiarian skill which, though it may easily be attained, is greater than is usually possessed.

## II. MODERN BEE HIVES.

### NUTT'S COLLATERAL HIVE. No. 1.

The late Mr. Nutt, author of " Humanity to Honey Bees," may be regarded as a pioneer of modern apiarians; we therefore select his hive wherewith to begin a description of those we have confidence in recommending. Besides, an account of Mr. Nutt's hive will necessarily include references to the various principles which subsequent inventors have kept in view.

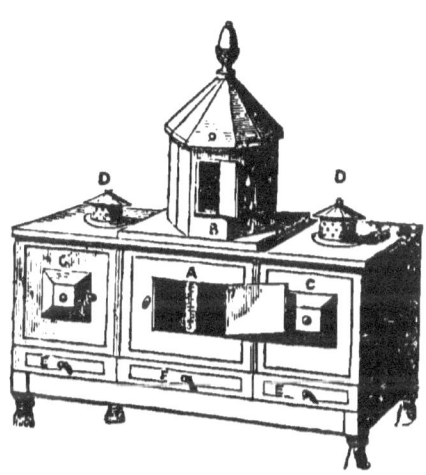

Nutt's Collateral Hive consists of three boxes placed side by side (C. A. C.), with an octagonal box B on the top which covers a bell-glass. Each of the three boxes is 9 inches high, 9 inches wide, and 11 inches from back to front; thin wooden partitions,—in which six or seven openings corresponding with each other are made— divide these compartments, so that free access from one box to the other is afforded to the bees; this communication is stopped when necessary by a zinc slide passing down between each box. The octagonal cover B is about 10 inches in diameter and 20 high, including the sloping octagonal roof, surmounted with an acorn as a finish. There are two large windows in each of the end boxes, and one smaller one in the centre box; across the latter is a thermometer scaled and marked, so as to be an easy guide to the bee-master, showing him by the rise in temperature the increased accommodation required. This thermometer is a fixture, the indicating part being protected by two pieces of glass, to prevent the bees from coming between it and the window, and thereby obstructing the view.

D D are ventilators. In the centre of each of the end boxes is a double zinc tube reaching down a little below the middle, the outer tube is a casing of plain zinc, with holes about a quarter of an inch wide dispersed over it; the inside one is of perforated zinc, with openings so small as to prevent the escape of the bees, a flange or rim keeps the tubes suspended through a hole made to receive it. The object in having double tubing, is to allow the inner one to be drawn up and the perforations to be opened by pricking out the wax, or rather the propolis, with which bees close all openings in their hives. These tubes admit a thermometer enclosed in a cylindrical glass, to be occasionally inserted during the gathering season; it requires to be left in the tube for about a quarter of an hour; and on its withdrawal, if found indicating 90 degrees or more, ventilation must be adopted to lower the temperature—the ornamental zinc top D must be left raised, and is easily kept in that position by putting the perforated part a little on one side.

The boxes before described are placed on a raised double floor-board, extending the whole length, viz., about 36 inches. The floor-board projects a few inches in front. In the centre is the entrance;—as our engraving only shows the back of the hive, we must imagine it on the other side,—it is made by cutting a sunken way of about half-an-inch deep and 3 inches wide, in the floor-board communicating only with the middle box; it is through this entrance alone that the bees find their way into the hive, —access to the end boxes and the super being obtained from the inside. An alighting board is fitted close under the entrance for the bees to settle upon when returning laden with honey; this alighting board is removable for the convenience of packing. The centre, or stock-box, A, called by Mr. Nutt the *Pavilion of Nature*, is the receptacle for the swarm; for stocking this, it will be necessary to tack the side tins so as to close the side openings in the partition, and to tack some perforated zinc over the holes at top; the swarm may then be hived into it just the same as with a common hive. A temporary bottom-board may be used if the box has to be sent any distance; or a cloth may be tied round to close the bottom (the latter plan is best, because allowing plenty of air), and when brought home at night, the bees being clustered at the top, the cloth or temporary bottom must be removed, and the box gently placed on its own floor-board, and the hive set in the place it is permanently to occupy. E E are two block fronts which open with a hinge, a semicircular hole 3 inches long, 2 wide in the middle, is cut in the upper bottom-board immediately under the window of each box; these apertures are closed by separate perforated zinc slides; these blocks, when opened, afford a ready means of reducing the temperature of the side boxes, a current of air being quickly obtained, and are also useful for allowing the bees to throw out any refuse.

The centre F is a drawer in which is a feeding trough, so constructed that the bees can descend through the opening before mentioned on to a false bottom of perforated zinc; liquid food is readily poured in by pulling out the drawer a little way, the bees come down on to the perforated zinc and take the food by inserting their proboscis through the perforations, with no danger of being drowned. Care must be exercised that the food is not given in such quantity as to come above the holes; by this means, each hive has a supply of food accessible only to the inmates, with no possibility, when closely shut in, of attracting robber bees from other hives.

The exterior of these hives is *well* painted with two coats of lead colour, covered with two coats of green, and varnished. Notwithstanding this preservation, it is absolutely essential to place such a hive under a shed or cover of some sort, as the action of the sun and rain is likely to cause the wood to decay, whilst the extreme heat of a summer sun might cause the combs to fall from their foundations.

Neat and tasteful sheds may be erected, either of zinc supported by iron or wooden rods, or a thatched roof may be supported in the same manner, and will form a pretty addition to the flower garden.

When erecting a covering, it will be well to make it a foot or two longer, so as to allow of a cottage hive on either side, as the appearance of the whole is much improved by such an arrangement.

The following directions, with some adaptation, are from "Nutt on Honey Bees:"—

In the middle box the bees are to be first placed;—in it they should first construct their beautiful combs, and under the government of one sovereign—the mother of the hive—carry on their curious work, and display their astonishing architectural ingenuity. In this box, the *regina* of the colony, surrounded by her industrious, happy, humming subjects, carries on the propagation of her species, deposits in the cells prepared for the purpose by the other bees, thousands of eggs, though she seldom deposits more than one egg in a cell at a time : these eggs are nursed up into a numerous progeny by the other inhabitants of the hive. It is at this time, when hundreds of young bees are daily coming into existence, that the collateral boxes are of the utmost importance—both to the bees domiciled in them, and to their proprietors; for when the brood become perfect bees in a common cottager's hive, a swarm is the necessary consequence. The queen, accompanied by a vast number of her subjects, leaves the colony, and seeks some other place in which to carry on the work nature has assigned her. But as swarming may by proper precaution and attention to this mode of management generally be prevented, it is good practice to do so; because the time necessarily required to establish a new colony, even supposing the cottager succeeds in saving the swarm, would otherwise be employed in collecting honey, and in enriching the old hive. Here, then, is one of the features of this plan—viz., *the prevention of swarming.* When symptoms of swarming begin to present themselves, which may be known by an unusual noise, the appearance of more than common activity among the bees in the middle box, and, above all, by a sudden rise of temperature, which will be indicated by the quicksilver in the thermometer rising to 75 degrees as scaled on the thermometer in the box; when these symptoms are apparent, the bee master may conclude that additional space is required. The top sliding tin should now be withdrawn from under the bell glass, which will open to the bees a new store-room; this they will soon occupy, and fill with combs and honey of pure whiteness, if the weather be fine to allow of their uninterrupted labour. It may be well here to mention, that if the glass have a small piece of clean worker comb attached to the perforated ventilating tube, the bees will more speedily commence their operations in it. When the glass is nearly filled, which in a good season will be in a very short space of time, the bees will again require increased accommodation; this will also be indicated by the thermometer further rising to 85; the end box, as thereon marked, must now be given them. Previously to drawing up a slide to enlarge their crowded house, the manager should take off the empty end box he intends to open to them, carefully and thoroughly cleanse it, and then smear or dress the inside of it with a little liquid honey. Thus prepared, he must return the box to its proper situation, and then withdraw the sliding tin that hitherto has cut it off from the middle box; by so doing the store-room is again enlarged. The bees will commence operations in this new apartment. This simple operation, done at the proper time, generally prevents swarming; by it, the queen gains a vast addition to her dominions, and, consequently, increasing space for the multiplying population of her domicile. Provided the weather continue fine, and the thermometer has risen to 95 degrees, as marked on the scale, the remaining tin may be also withdrawn, thereby giving the bees, admittance to another box; there is now no lack of store-rooms nor of employment for our indefatigable labourers. The cylinder thermometer is required to be occasionally dropped into the ventilating tube of the side boxes to ascertain their temperature; for if exceeding or approaching that of the middle box, it must be reduced by ventilating; this is done by raising the zinc tops, to allow the air to pass through the perforations. The grand object of this system is to keep the end boxes and the bell glass cooler than the pavilion or middle box, so as to induce the queen to propagate her species there and there only, and not in the depriving part of the hive; by this means the side and upper combs are in no way discoloured by brood. The queen requires a considerable degree of warmth; the middle box does not require more ventilation than the additional openings afford. The bees enjoy coolness in the side boxes, and thereby the whiteness and purity of the luscious store are increased.

After having given directions for the working of the hive, it remains to be told how to obtain possession of the store, and to get rid of our industrious tenants from the super and end boxes, of which the super glass will be almost sure to be filled first, having been first given to them. The operation of taking honey is best performed in the middle of a fine sunny day. The best mode that we know of is to pass an ordinary table-knife all round underneath the rim of the glass to loosen the cement, properly called propolis; then take a piece of fine wire, or a piece of string will do, and, having hold of the two ends, draw it under the glass very slowly, so as to allow the bees to get out of the way. Having brought the string through, the glass is now separated from the hive; but it is well to leave the glass in its place for an hour or so, the commotion of the bees will then have subsided; and another advantage we find is, that the bees suck up the liquid and seal up the cells broken by the cutting off. You can then pass underneath the glass two pieces of tin or zinc; the one may be the proper slide to prevent the inmates of the hive coming out at the apertures, the other tin keeps all the bees in the glass close prisoners. After having been so kept a short time, the apiarian must see whether the bees in the glass manifest symptoms of uneasiness, because if they do not, it may be concluded that the queen is among them. In such a case, replace the glass, and recommence the operation on a future day. It is not often that her majesty is in the depriving hive or glass; but this circumstance does sometimes happen, and the removal at such a time must be avoided. When the bees that are prisoners run about in great confusion and restlessness, the operator may then conclude that the queen is absent, and that all is right. The glass may be taken away a little distance off, and placed in a flower-pot or other receptacle where it will be safe when inverted and the tin taken away, then the bees will be glad to make their escape back to their hive. A little tapping at the sides of the glass will render their tarriance uncomfortable, and the glass may then be taken into a darkened room or out-house with only a small aperture admitting light which must be open; the bees, like all insects, make towards the light and so escape. The bee-master should brush them off with a feather from the comb as they can be reached; but on no account, if there are many bees, should the glass be left, because the bees that are in the glass will gorge themselves to their full and speedily bring a host of others from the adjacent hives, who, in a very little time, would leave only the empty combs. It is truly marvellous how soon they will carry all the store back again, if allowed to do so. An empty glass should be put on to the hive in place of the full one, as it will attract the bees up, thereby preventing the too close crowding of the hive; and, if the summer be not too far advanced, they will work more honey-comb in it.

The taking away of the end boxes is a somewhat similar process; but they should on no account be taken at the same time as the glass, or indeed at the same time as any other hive may be—*robbed* we were going to say, for it is robbery to the bees,—they intended the honey for their winter food, and are much enraged at being deprived of it. First shut down the dividing tin; the bees in the end box are now prisoners separated from the hive, keep them so half an hour, then take away the box bodily to another part of the garden, or into the dark out-house as before recommended.

It may not be out of place here to say something respecting the enthusiastic inventor of the Collateral Hive—Thomas Nutt, who was an inhabitant of Spalding, in Lincolnshire. Having been disabled during a considerable period by rheumatic fever, he devoted all his attention to bees, at a time when bee-culture was but little valued; and, although it must be admitted that two boxes were used side by side long before Mr. Nutt's day, still it is due to him to state that the adoption of three boxes was entirely his own idea, and that as far as he then knew, the collateral system was his original invention. His statements have been severely criticised, and it does appear that the weight of honey which he names as having been produced in one season is perfectly incredible. But as in the district where he lived there is grown an immense quantity of mustard seed—the flowers of which afford excellent forage for bees—the honey harvests there, would doubtless, be very large. If Mr. Nutt has given his little favourites too much praise, it will be only charitable, now, to account for his statements by an excess of zeal and enthusiasm in this his study of bee-culture. It may be that the golden harvests he spoke and wrote of have been so far useful that they have induced many to commence bee-keeping, some of whom, whilst they condemned his statements, have themselves written really useful and practical works on the subject, which otherwise might possibly never have appeared. As the monks of old kept the lamp of religion burning, however dimly, until a more enlightened age, so Thomas Nutt may have assisted in a somewhat similar manner by energetically propounding his views, and thereby causing other apiarians to rise up whose names are now as familiar to us as household words, and whose works posterity will value. The writer of these pages has often accompanied Mr. Nutt on his visits to his patrons in the neighbourhood of London, and seen him perform his operations regardless of the anger of bees, and free from all fear of their stings. He often expatiated on the cruelty of the brimstone match and suffocation, denouncing the barbarous custom in the following terms:—"You may as well kill the cow for her milk, or the hen for her eggs, as the bee for its honey; why continue to light the fatal match, when every cottage in England has the means of saving this most useful and valuable insect?"

## NEIGHBOUR'S IMPROVED SINGLE BOX HIVE. No. 2.

We have introduced the "Single Box Hive" to suit the convenience of those who, though desirous of keeping bees on the improved principle, do not wish to incur the expense or devote the space which is necessary for Nutt's hive.

It consists of a lower or stock-box A., 11 inches square, 9 inches deep, with three large windows, a thermometer D, as in Nutt's, being fixed across the front one, protected at the sides by strips of glass to prevent the bees obscuring the quicksilver from sight. B is a cover the same size as the lower hive, large enough to allow space for a bell-glass 9 inches wide, 6 inches deep. E is the ventilator between the glass and the stock-hive, intended to prevent the queen travelling into the super hive, and also by cooling the hive to endeavour to prevent swarming; a sloping pagoda roof with an acorn top completes the upper story. A floor-board with a block front, as in Nutt's collateral, forms the base, the entrance being sunk as before described, and furnished with zinc slides to reduce or close it as may

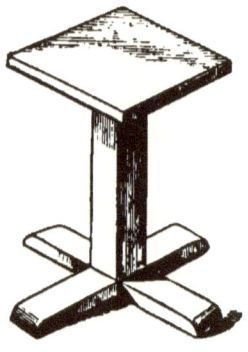

be required. To stock a hive of this description, it is necessary to send the stock-box to the party with whom you have agreed for the supply of a swarm. In the evening of the day the hive is thus tenanted, remove it to the position it is designed permanently to occupy; if the swarm has to be procured from a distance, and is transported by rail or other conveyance, a perforated zinc slide should be substituted for the plain slide that covers the top, and a large piece of perforated zinc must also be tacked to the bottom after the swarm has settled in; thus securely confined, with a free circulation of air throughout, bees that have been swarmed the day before may be safely sent any distance that will allow of their being released the day after, because bees, though they provision themselves for a couple of days, cannot with safety be confined in an empty hive much longer.

Having now, we will suppose, procured your swarm, and having placed it in a south or south-east aspect, you may with advantage, if the weather be wet, give a little liquid food ; the feeding in this hive is performed at the top of stock-box, where the glass is worked. Our round Feeding Pan or the new Feeding-Bottle may here be used. Any fancy as to the position may be indulged in, but must be settled on by the time the bees are set at liberty, because any alteration afterwards is detrimental to the working of the hive. The bees on first issuing forth, carefully mark their new abode and the surrounding objects, so that if a change be made, they are completely thrown out in their observations, which confuses them not a little, and occasions loss. Bees always return to the same spot; it is the locality that they know, and if the hive is moved a less distance than a mile, thousands return to the spot on which the hive has been accustomed to stand.

Allow your bees to collect honey and build their combs for ten days or a fortnight. Much now depends on the weather; if fine, by this time they will require additional room, which will be indicated by the thermometer D rapidly rising; 100 degrees is the swarming point, the hive must be kept below this by ventilation.

Access must now be given to the flat bell-glass at the top, which is done by withdrawing the top slide. In a few hours, sometimes immediately, the work of comb-building begins in the glass—all the sooner, if a piece of clean empty comb be placed therein.

It is of service to keep the glass warm by means of a worsted or baize bag, it prevents the temperature from falling at night when much comb-building is carried on, providing the heat is not allowed to escape. Probably, if all goes on well, in three weeks the glass will be found to be filled with fine white honeycomb. When you find that the comb is well sealed up, it is time to take it off, but if the cells are unfilled and unsealed, let the little labourers complete their work—a little experience will soon enable the bee-keeper to determine this point.

The plan to be adopted for taking glasses of honeycomb is the same as described for hive No. 1, page 21.

## TAYLOR'S AMATEUR SHALLOW BOX OR
## EIGHT BAR HIVE. No. 3.

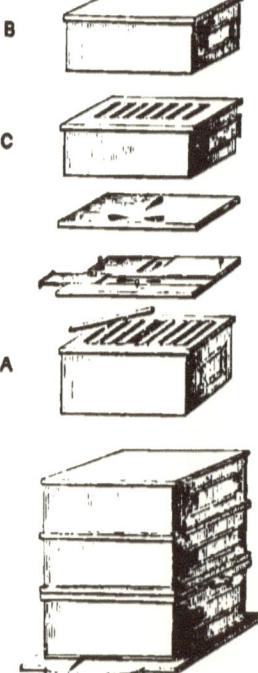

Taylor's Amateur Hive, as seen by reference to the engraving, consists of three boxes, the lower one A, is the stock box, in which the swarm is first placed; B is the first super; and C the centre box; all three boxes are of the same diameter, viz., 13½ inches square inside; A, the stock box, is 7½ deep; B, 6½ inches; both are fitted with eight moveable bars, each bar being 1⅛ wide, with spaces of half an inch between, and all are easily removed by unscrewing the crown board, in which are two openings closed by zinc slides; the middle box C has no bars, and is still shallower than either of the other boxes, being 5 inches deep. In many localities and seasons, the third box may not be required. Each box has two windows —one at the back and another at the side, a zinc shutter sliding in a groove, excluding light and retaining warmth; the box C differs from the others in another respect; instead of bars it has a grating made by seven openings, each ½ an inch wide and 9 inches long; these three boxes stand on a stout floor-board, in which is cut the entrance way, 4 inches wide and ¾ of an inch high. The floor-board projects so as to support an outer cover of half-inch wood, surmounted by a sloping roof. This is an effectual protection from the weather, and is necessary when hives are exposed; of course, if placed in a bee-house such protection may be dispensed with. The outer case is well painted, of a green colour, and when it is used the hive may be placed in any part of the garden. The dimensions of this hive, with outside cover, are 18 inches square, 2 feet 6 inches high.

Suitable stands are provided, consisting of a stout pedestal with four feet. Stakes should be driven into the ground to secure the whole against wind. Height from the ground, 4 feet 8 inches.

The bars before alluded to are for the purpose of inducing the bees to build parallel combs; for without this, extraction would be impossible. It is a great convenience, in many ways, to be able to take out a bar of comb, it gives such complete control over the hive.

To ensure comb-building on the bars, pieces of clean worker comb should always be carefully preserved; and before a swarm is put in, either every bar, or if guide comb is not plentiful, every other bar should have a piece fixed to it in the following manner: cut a piece of clean empty comb of the required size, say two inches square, not less; heat a common flat iron, and slightly warm the bar with it, then melt a little bees-wax upon it ; draw the comb quick over the heated iron, hold it down on the centre of the bar, giving a very slight movement backwards and forwards, then leave it to grow cold; and if cleverly managed, it will be found to be firmly attached. Care must be taken that the pitch or inclination of the comb be the same as it is in the hives—upwards from the centre of each comb. A new plan has lately been introduced by Mr. Woodbury, of Exeter, to facilitate the correct construction of parallel combs.

## TAYLOR'S AMATEUR BAR HIVE. No. 4.

Taylor's Amateur Bar Hive is stocked exactly in the same way as before described—viz., by hiving the swarm into the lowest box A, as with an ordinary Cottage Hive, and in a fortnight's time the box B is placed over it, and the zinc slides withdrawn. After this has been given them, and is nearly filled, the super B is raised, and the box C is placed between, immediately over the stock box, to induce the bees to continue the combs.

This hive consists of three boxes, one above the other, similar to the No. 3. The boxes are less in diameter than the foregoing, and have seven moveable bars in each. Recent improvements and observations led Mr. Taylor to prefer eight bars, and to have the boxes made a little more shallow.

The mode of stocking and management of this hive are the same as that last described. There is no outside cover or protection from the weather, but the wood is additionally thick and is well painted.

We quite concur with Mr. Taylor in recommending a broader and shallower hive, and advise intending purchasers to select the eight-bar hive in preference.

### NEIGHBOUR'S IMPROVED COTTAGE HIVE. No. 5.

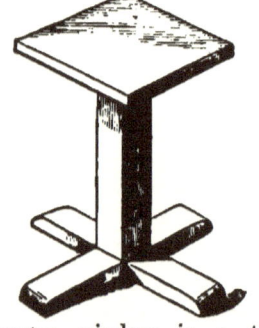

Our "Improved Cottage Hive" is neatly made of straw bound with cane, and therefore very durable.* The lower hive is covered with a wooden top having in it three holes, through which the bees convey their honey into three middle sized bell glasses with ventilators, which when filled hold about 6 lb. each. There is a hoop at the bottom, another round the top of the lower hive; to this the wooden crown board is fastened. These hoops are a great improvement, and are less liable to harbour insects than if straw alone were used. The floor-board, as its name implies, is a wooden board, 1¼ inch thick, with a projection of 3 or 4 inches under the entrance to form an alighting place. This entrance is cut out of, or sunk in the board.

There are three windows in the lower hive, each closed with a shutter, these are very useful and interesting for inspecting the progress made. Across the centre window is a thermometer, enclosed at the sides by

---

* This is the hive referred to by the Bee-Master of the *Times*, when he says:—"The second kind of hive I alluded to is made of straw, and may be purchased at Neighbour's, in Holborn. * * * * It is so well made that it will last very long. I have had one in constant use during ten years, and it is still as good as when it was bought."

slips of glass. The window shutters being painted green, add very much to its appearance. The upper hive, which is merely a cover for the glasses, is a conical topped hive, also made of straw bound with cane; a hoop is worked into the straw, and made sufficiently large to allow the cover to drop over the top hoop of the lower hive, keeping the whole close, and preventing wet from drifting in. A zinc ventilator, ornamentally painted, forms the apex: this is useful in letting the confined hot air pass away in warm weather. The ventilator is opened by raising it. The dimensions of the lower or stock hive are 15 inches diameter, 9½ inches deep outside, its weight when empty 7½ lbs., the cover or top hive is 12 inches deep, and 15 inches in diameter, the ornamental zinc top being 4 inches deep. The whole is about 24 inches high. The weight of a hive packed, including glasses, &c., is about 18 lbs.

These hives have a tasteful appearance in the garden, but they require some further protection from the weather in the form of a cover or of a bee-house—contrivances that have yet to be described. In extreme cold weather, a little additional protection by having matting folded round them will be advisable.

One of the advantages this hive has over the common cottage hive is that it affords opportunity for the humane management of bees. The owner has also the power of taking a glass of honeycomb of pure quality, free from the extraneous matter, known as " bee-bread," instead of combs that are darkened by having brood hatched in them. By this system, we have combs newly made and used only for depositing the honey first put into them, hence the name " virgin honey." These glassss have a very pretty appearance, and when nicely filled, are very convenient for home use or for making presents. The lower hive is the receptacle for the bees; when a swarm is placed in this hive, they immediately proceed to fill it with combs, in which to store honey for themselves, and for cells to breed in. This hive remains undisturbed.

The best mode of tenanting a hive of this description is by placing an early and strong swarm in it, which may be generally procured of a neighbouring bee-keeper; if from a distance, considerable care is necessary to admit plenty of air; the shaking attendant upon carriage irritates the bees so much that, if not well ventilated, there is danger of the swarm being stifled, and the finer the swarm, the greater the danger. For the purpose of ventilation, remove the slides and substitute perforated zinc, wrapping the hive up in a coarse cloth of open texture (dispensing with the floor-board during transit when the distance is great).

It is necessary only to send the lower or stock hive to the party furnishing the swarm, taking the precaution to fix the slides at top with tacks, as the hive has to be inverted to receive the bees. They are shaken into it in the usual manner, as they cluster around the branch of the tree or shrub on which they may have chosen to alight. After the hiving is accomplished, the hive should be left near to catch any stragglers, for there will always be a few; towards evening, close the entrance, and remove them to the exact position they are intended permanently to occupy. Success depends on this, and also on their careful removal on the day or evening of swarming. The following morning the bees labour in the new location, marking well their habitation before they take flight, and to which they will not fail to return loaded with luscious store.

A fortnight must be allowed for filling the stock hive; then, if the weather be fine and warm, they will prepare to swarm again, as will be indicated by the thermometer rising rapidly to 100 degrees or upwards; one of the zinc slides on the wooden top must now be withdrawn, and a bell glass put on covered and protected by the upper hive, the other glasses may then be given in the same manner; a day or two after which, should the weather continue favourable, all signs of swarming will at once disappear, the bees now having increased store room which they will readily fill with comb. It is often found useful to attach a piece of clean empty honey comb to the ventilating tube of the glass; it is an attraction, and induces the bees to commence working in it sooner than they otherwise would do. The ventilator should also remain open during the day to allow the hot air to pass away from the interior, thereby contributing to the whiteness and beauty of the work; the bees enjoy the refreshment of coolness thereby afforded, and they work the faster for it. At evening all ventilation should be stopped, and the glasses wrapped round with flannel or some warm material, for the reasons mentioned on last page.

The directions for taking honey are also the same as before mentioned.

The holes on the wooden top of this hive are of a peaked shape, being a preventive against slaughtering any bees whilst pushing the slide in for the purpose of removing the glass when full; the tacks before alluded to should be removed from the slides when the hive is fixed in its place, they are now in the way of cutting off the glass. The hive entrance has two slides: the perforated one is but seldom required, the bees object to being closely confined, it is only necessary when removing, and then for as little time as possible; the other slide is very serviceable during the winter months to lessen the passage way, thereby preventing the admission of too much cold air; it is also occasionally useful on a summer evening, to lessen the entrance when moths are troublesome, for if there be only a small opening, the bees can guard it, and easily repulse intruders. During the time of gathering they require the whole width to remain open.

When the weather is so unfavourable as to prevent the bees leaving home for a few days after being hived, it will be necessary to feed them. Bees should not be fed in the midst of winter; the proper time is in the autumn or in the spring.

The best mode of feeding is at the top of the Stock Hive. This is done by using the Round Feeder.

The Bottle Feeder may be used instead of the Round Feeder, and in the same place, by those who give the preference to that method.

The simplicity and easy management of this hive have deservedly rendered it an especial favourite, combining, as it does, real utility with many conveniences to satisfy the curious. Not a few bee-keepers desire to unite the two qualifications, and no hive combines these advantages in a greater degree than "Neighbour's Improved Cottage Hive."

## IMPROVED COTTAGE HIVE. No. 6.

The No. 6 hive is of precisely the same size, construction, and management as the last mentioned, with the exception that it has no windows or thermometer in the lower or stock hive. The apiarian with this hive will have to trust more to his own judgment as regards the likelihood of swarming, and must watch the appearance the bees present at the entrance. When it is time to put on supers in order to prevent swarming, premonition will be given by the unusual numbers crowding about the entrance, as well as by the heat of the weather, making it evident that more room is required for the increasing population.

Not being able to form an idea of the state of the hive in spring and autumn by looking into the stock hive, it will be advisable to adopt the means of weighing. A stock at Michaelmas should weigh 20 lbs. exclusive of the hive, or be made up to that weight by feeding.

## THE LADIES' OBSERVATORY OR CRYSTAL BEE-HIVE. No. 7.

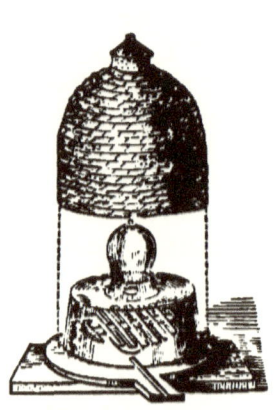

This engraving illustrates the construction of the "Ladies Observatory" hive. The stock hive cylindrical with a flat top and a hole in the centre; the dimensions 12½ inches inside, 8½ inches deep, the outer cover being raised, and is made of stout glass, so as not easily to break.

A support, composed of seven wooden bars fixed on a pedestal from the floor-board, is very useful for the bees to cling to and attach their combs, instead of resting wholly against the glass.

The floor-board is of mahogany, the border being French polished. A middle sized bell-glass for deprivation is placed over the hole; this hole may be closed by a zinc slide. A cover of straw, 18 inches deep, 15 inches wide, with a zinc ventilating top similar to that affixed to the No. 5 hive, completes the arrangements. The weight of the stock hive and board is about 16 lbs.

This hive is well adapted for those persons who are desirous of having the opportunity of more closely examining the workmanship of these industrious and interesting insects, as the whole of the interior may be exposed to view; it is particularly suitable for a window or an indoor apiary, and also will be found a valuable addition to the green house. Under these circumstances, the entrance-way should be covered with a flat piece of glass, and an aperture cut in the sash corresponding with the entrance to the hive; through the glazed passage the bees may then find egress and ingress without being able to gain access to the apartment; an alighting board four inches wide must be fixed outside on a level with the entrance.

We had a hive of this kind in operation in the Great Exhibitions of 1851 and 1862, fixed in the manner above described. It answered admirably, and excited much interest and curiosity, though placed there under many disadvantages.

When a hive of this kind is to be stocked, procure an early and strong swarm, which must be temporarily hived in a common straw hive, from which dislodge the bees into the glass hive, but for this purpose a little preparation will have to be made. Spread a sheet on the ground, place the mahogany floorboard on it with the support, put three bricks, or some solid blocks of about the same substance, upon which the glass will rest; then with a sharp and sudden blow precipitate the swarm out of the straw hive on to the floorboard and support, place the glass hive on the bricks, and the bees will collect under the bars and on to the pedestal. In about one hour's time the whole will have settled quietly, and all the stragglers on the board will have collected together, the swarm hanging pear-shaped from the bar support; the bricks can now be removed, and the glass put in its right place on the floorboard. The straw cover being put on the hive, it can be removed to the place it is destined permanently to occupy.

The light should not be admitted for some days after hiving; if undisturbed, the bees will speedily build comb, working from the wooden bars which are placed there for their assistance and support. In ten days or a fortnight, if the weather continues fine and warm, they will prepare to swarm again; the opening at the top must now be unstopped, and the bell-glass put on, guide comb having been previously fixed. The directions given for the No. 5, Cottage Hive equally apply to the Ladies' Observatory Hive.

It is advisable in winter to furnish the glass stock hive with more protection from cold than is afforded by the straw cover alone; some thick baize or wrapper of wadding, for which there is space between the glass hive and the cover, will prevent so much moisture condensing on the sides of the glass. Moisture is injurious, causing the combs to grow mouldy; a little protection in the way of wrapping very much prevents this.

The hole at top is used for supplying food should the apiarian fear the stock of honey is in danger of running short; either the bottle feeder, or the round feeder, may be used for the purpose.

D

## COTTAGER'S HIVE, No. 8.

### FOR TAKING HONEY IN STRAW CAPS WITHOUT THE DESTRUCTION OF THE BEES.

A very prevalent opinion exists that bees do better in straw than in hives made of any other material. Another opinion prevails, viz., that the old fashioned straw hive is the least expensive, the most simple, and the most productive. Although we cannot go so far as this, we are willing to admit that a simplified adaptation of the humane system to the old common straw hive is the most suitable to put into the hands of that large class of bee-keepers— *Cottagers.* By these the more fanciful hives will be instantly condemned; besides, the expense puts them quite beyond the reach of the poorer class. The object aimed at in planning our Cottager's Hive has been to furnish a depriving hive that should be at once easy of management, in-expensive, and convenient. The stock hive, into which the bees are first hived, is a round straw hive, having a flat top with a hole in the centre. The size of this lower hive is **7** or **8** inches deep, 14 inches across the bottom, finished with a wooden hoop, which adds very much to the firmness and durability of the hive. The floorboard is 1¼ inch thick, with a way sunk therein for the entrance. A small round mat of straw closes the hole on the top; this mat may be fixed by wooden pegs. We have now described what is termed the *stock hive,* which is, in fact, an old fashioned straw hive adapted, modernised, and improved to the more humane, viz., the depriving system. The weight of the stock hive with its floorboard is about 7 lbs.

The super or cap hive is about 7 inches deep, 8 inches in diameter, and when filled contains about 10 lbs. of honey and comb. A glass window which is placed at the side is useful for inspecting the progress made in filling it.

A common straw hive, sufficiently deep to cover, drops over the super, keeping the window dark and fitting close on to the stock hive.

This cover hive may be made fast by driving in two skewers, one on either side, to keep the whole firm. Unless placed in a bee-house or under a shed, the outside should be painted; or a piece of oil-cloth or waterproof covering of any kind shaped so as to shoot off the rain, will save the trouble of paint, and answer the purpose. If no protection of this sort is used, the rain is likely to rot the straw. As a covering cottagers often use straight stiff thatching straw sewed together; this contrivance is termed a hackle, and has a pretty appearance, particularly if a number of hives are in a row. Care has to be exercised that mice do not make the covering hive a resting place. Mortar is often used for fastening round the hive at the bottom; this is a bad plan, as it forms a harbour for insects; the wooden hoop fits so close as to leave little necessity for anything of the kind.

The principle of the depriving system is so much the same with all our hives, that a good deal of repetition is necessary in describing in detail the management of each separate variety. The object aimed at with the Cottager's Hive, as indeed with all our hives, is to provide a compartment for the bees to live in with their queen, she being the mother of all. It is intended, by inducing the queen to remain in her original apartment, that all breeding should be there performed, as well as the storing of bee-bread and honey for the winter sustenance of the bees. The cap hive or upper chamber, known as the "super," is for the storing of honey which the bee-keeper looks upon as a surplus, and which, at the close of the honey gathering, or as soon as filled, he intends to deprive the bees of, and appropriate to his own use,—of course taking care to leave sufficient in the lower or stock hive for winter sustenance.

The mode of stocking a hive of this kind is so familiarly known that any who at all understand the hiving of bees into a common straw hive, can make no mistake or find any difficulty in performing it. Lest these pages should fall into the hands of persons who are not so acquainted, we will refer them to the directions already given.

The hive may be smeared inside with a little honey if at hand; but this is unimportant, as a clean hive answers well. Some older bee-keepers prefer to give a little dressing to encourage the bees to like their new home.

After the swarm has been in the hive two weeks, the straw super hive may be put on, first removing the straw mat to give the bees access to it. If the hive be a stock, that is a swarm of the last or previous years, the super may be put on as soon as the weather is fine and warm in May. But much depends on the weather and the strength of the hive as regards the time occupied by the bees in filling the super; in favourable weather a fortnight suffices.

If, on looking in at the little window, the bee-master sees that the cells are sealed over, the cap of honey may be removed in the mode already described. The cells near the window are the last to be filled, so when they are sealed, it is safe to conclude that the combs in the unseen parts are also finished.

Sometimes the queen ascends and deposits her eggs; if, on turning up the super, brood be visible, replace the cap for a few days, until the young bees quit their cells. When thus emptied, honey will be deposited in lieu of the brood.

Suitable pedestals for these hives to stand upon may be obtained. It is important that these be firmly fixed, and the hive also made fast to the stand, to prevent its being blown over by high winds.

## BAR AND FRAME HIVES.

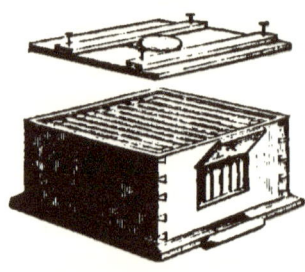

Mr. Woodbury's Bar and Frame Hive, as originally made, consists of a wooden box 14½ inches square inside, 9 inches deep. This is a hive of large size, but the actual habitable space inside is lessened by the room occupied by the frames, of which there are ten; these rest on a rabbet a little below the surface, leaving a space of ¾ of an inch between the upper side of the bars and the crown board. This allows a free passage on the top for the bees, entirely obviating the necessity of making excavations in the crown board, as has hitherto been recommended. Each frame is ⅞ of an inch wide, and rests in notches, with a space of ½ an inch between each. The frames extend to within ¾ of an inch of the floorboard, so as to hang without touching any part, leaving about the same distance from the sides. It will be seen that there is a free passage for the bees on every side, and they are thus kept from coming in contact with the sides of the hive. Our engraving shows the hive open, and exposes to view the top of the ten bars and frames as they range from back to front. A window is also shown; this is placed in the engraving over the entrance, but the proper position would be just opposite. The drawing is made so as to show back and front at once. The floorboard is 1¼ inch thick, having two "keys" on the underside to prevent warping.

## STRAW BAR AND FRAME HIVE. No. 45.

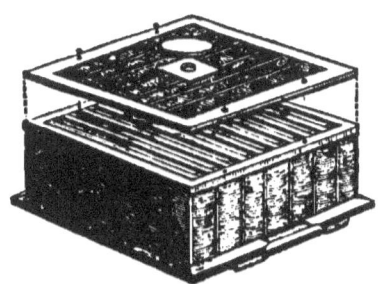

Since the introduction of the wood hive by Mr. Woodbury, that gentleman has recommended, in the *Journal of Horticulture*, that the stock hive be made of straw, of exactly the same dimensions; this material being warmer in winter, slightly ventilating, and allowing of absorption. Bees during cold weather cluster together to generate the requisite degree of heat, the temperature of the interior of the hive being thus so much higher than the external atmosphere, a good deal of moisture condenses at the top and on the sides of the hive; the straw, as before stated, prevents this dampness hanging about the hive, and tends to keep the inmates more healthy. Dampness in a hive is a fruitful source of mischief, causing empty combs to grow mouldy, and is injurious in many ways.

The square straw hives, and a machine for making them, exhibited in the Austrian department of the International Exhibition of 1862, suggested the idea of employing that material for English bar and frame hives. We have had made a machine somewhat similar to the one exhibited and suited to the size of our hives, by which our hive maker is able to manufacture neat square straw hives. These have a wood frame at top an inch deep, with the requisite notches to allow the ten comb frames to hang. A similar frame forms the base, the straw being worked between. The floorboard is 1¼ inch thick, "keyed" with stout keys as before mentioned. An inch projection is left on all sides beyond the exterior of the hive, from which it is slightly chamfered down. An entrance 4 inches wide is cut out of the substance of the board, beginning at the edge, and continuing on the same level until inside the hive, where it slopes upwards. This entrance is about ⅜ of an inch high where the hive crosses it.

These straw hives have been generally made without windows; as Mr. Woodbury and other scientific apiarians so prefer them. They consider that glass windows are unsuited for winter, because then moisture condenses on the glass. There is no doubt that the having a peep hole or two in a hive adds very agreeably to its value for amateur bee-keepers, and to meet the wishes of such, we have had straw hives constructed with windows. It is not every one who would like to lift out the frames as often as is necessary for an inspection of the state of the colony, nor, perhaps, is it advisable to be often thus meddling. The windows have also a very neat appearance. We have hives with one, and some with two and three windows; of course, a little extra expense is incurred where these are made; but that is not objected to by those who approve of the additional convenience. The crown board (if correct to call a straw top by that name) has, like the hive, a frame of wood all round, and a square piece of wood in the centre, with a two inch hole; this hole is for the purpose of administering food in a mode to be explained hereafter. A circular block of wood, 4 inches in diameter, closes the opening.

## GLASS BAR AND FRAME HIVES.

Some bee-keepers like to be able to make a full and daily inspection of the hive; we have therefore prepared a few hives, constructed of wooden frames enclosed on all sides and on the top with window glass. The dimensions are precisely the same as those before mentioned, and allow of the same number of bars and frames (ten). The crown has a round hole cut in the glass to admit of feeding. The four sides are constructed of double glass, to preserve the bees from variations of temperature. We cannot, however, recommend this hive for a winter residence for the bees; we should prefer lifting the combs out with the bees, and placing them in a straw hive of similar construction, to pass through the ordeal of the winter season. A stock of bees might be kept through the year in a hive of this kind, but would require well wrapping round to keep out the cold. There should be a small glass over the hole at top so as to allow the moisture to arise and condense, instead of doing so in the hive. The operation of exchanging the hive is so easy, that we should be content to place a stock in one, say, from April to September, and shift it in the autumn. Such a hive is a very pleasing object of interest, as in it the whole commonwealth of bees is exposed to view; and the hive need not be obscured from daylight, provided it be protected from sun and rain. All the external wood-work is of oak colour varnished, so that the appearance of the "Glass Bar and Frame Hive" is extremely neat and much approved of.

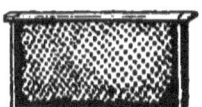 FRAME.—As before mentioned, each stock hive has ten of these frames, each 13 inches long by $7\frac{1}{4}$ inches high, with a $\frac{3}{8}$ inch projection at each upper end, which rests in the notch, either back or front. The width both of the bar and frame is $\frac{7}{8}$ of an inch; this is less by $\frac{1}{4}$ of an inch than the bar recommended by the older apiarians. Mr. Woodbury,—whose authority on the modern plans for keeping bees is of great weight,—finds the $\frac{7}{8}$ of an inch bar an improvement, because with them the combs are closer together, and require fewer bees to cover the brood. Then, too, in the same space that eight old fashioned bars occupied the narrower frames admit of an additional bar, so that, by using these, increased accommodation is afforded for breeding and the storing of honey.

 IMPROVED COMB BAR.—Mr. Woodbury says that this little contrivance has proved very effectual in securing straight combs when guide combs are not obtainable. The lower angles are rounded off, whilst a central rib is added of about $\frac{1}{8}$ of an inch in breadth and depth. This central rib extends to within $\frac{1}{2}$ an inch of each

Section of bar. end, where it is removed in order to admit of the bar fitting into the usual notch. All that is necessary to insure the regular formation of combs is, to coat the underneath surface of the central rib with melted wax. Mr. Woodbury further says, "my practice is to use plain bars whenever guide-combs are attainable, as these can be attached with much greater facility to a plain than to a ribbed bar; but whenever I put in a bar without comb, I always use one of the improved ones. By this method, crooked and irregular combs are altogether unknown in my apiary."

Most of our bars are made with the ridge; but should any of our customers prefer the flat ones, we keep a few to supply their requirements.

With the moveable "bar and frame hive" every comb is available for extraction, and may easily be taken out of the hive; each comb being fixed within its frame, there is less disturbance to the bees than if the combs were fixed to the sides, as is the case with ordinary hives. A strip of wood, about ½ an inch wide, rests on the floor-board; in this strip are ten notches, made to receive the lower part of the frames, so as to retain them in their places at equal distances from each other. A difficulty is found with a well stocked hive in dropping the frames into the exact notches, so that it is not necessary to have these rack works always in use; but when any movement of the hive is made, it is essential to have the frames firmly fixed by the aid of this contrivance. It is also advisable to have the frames perpendicularly supported until the combs are built, so in order that the frames should hang true, the hive ought to be on the level. A little inclination may be given to it from back to front, causing the hive slightly to fall towards the entrance, so as to allow the moisture inside the hive, caused by the exhalations of the bees, to run off.

"COMPOUND BAR FRAME."—In the *Journal of Horticulture*, Mr. Woodbury thus describes the compound bar frame. Being his own adaptation, we cannot do better than use his own words. "This is a contrivance of my own, which I have found very advantageous in enabling me to use frames in stock hives and bars in supers, without forfeiting the advantages arising from the unlimited interchangeability of every comb in every hive and super in the apiary. Its construction will be readily understood by an inspection of the annexed sketch, in which the comb bar is shown slightly raised from its frame. The bar itself is 13¼ inches long by ¾ of an inch wide and ⅜ of an inch thick. When the comb bar is in its place, the whole forms a frame 13 inches long by 7¼ inches high (inside measure), with ⅜ of an inch projection at each end, which rests in its appropriate notch in either the back or front of the hive.

When filled with comb, the bar becomes so firmly cemented to the frame as to admit of its being handled with facility." This contrivance is, no doubt, very excellent in the hands of Mr. Woodbury; but in the hands of the unpractised severe mishaps may arise. In warm weather the propolis and wax, with which the bees cement the bar to the frame, becomes soft, consequently in handling the frames, unless dexterity is used, the comb is likely to drop out. We, therefore, recommend that the bar and frame be made both in one: greater firmness and simplicity are thereby gained. Some of these compound bars and frames are kept in stock at our establishment, though they cannot be recommended for general use; but should any one prefer them, they can be supplied at the same price as the common frames.

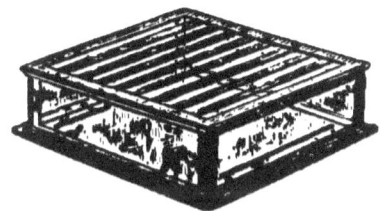

In describing the stock hives of wood, straw, and glass, allusion has frequently been made to the depriving hives, technically called " supers." These are also made of glass in wood frames, 13 inches inside, 6 inches deep, with eight bars (without frames). The accompanying cut represents the super used with the bar and frame hive.

Honey-combs in supers are better when made thicker than those for breeding, consequently the bars are placed a little further apart than in the lower or stock hive; they are either the " Woodbury Ribbed," contrived to induce the straight building of combs, or flat bars with guide combs affixed.

Cover.—A loose outer case, forming a complete cover for the hive, is found very useful. The case is made in two parts for convenience, the roof is also separate, having an acorn at top which forms a neat finish. These outside cases are made of wood, and drop lightly over all; when thus protected and fixed on a pedestal, the hive may be placed in the open air in such position as fancy may dictate. The aspect should be south or south-east, and if against a wall, sufficient space must be allowed for a free passage behind, as it is from thence all operations must be conducted by the apiarian. The case and roof with the stand being the only parts exposed to weather, will be the only portions that require painting; they are sometimes stained and

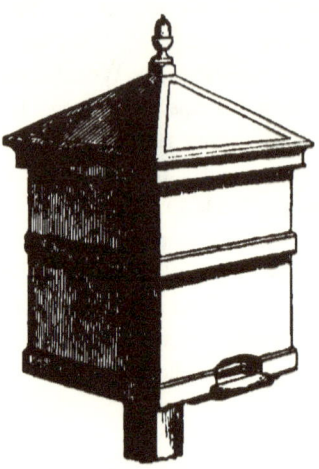

varnished, and we are inclined to prefer the latter for appearance. Should the apiarian have a complete bee-hive house, the cover and stands will not be needed.

When removing or replacing the covers, care should be taken to do so very gently, or the bees will be enraged and rush out, and may inflict stings upon those within their reach. The necessity for lifting off the cover for the purpose of looking in, either at the window of the stock hive or of the super, we have obviated by making a door both in the upper and lower parts of the outside case. These doors, or unglazed windows, are hinged at the bottom so as to open downwards, rendering inspection easy without disturbance to the bees.

A wooden range for supporting a number of hives makes a safe and economical stand; it may be formed by driving firmly into the ground two rows of posts, each row about 12 inches apart; to these two rails about 2 inches square are nailed, and upon these the hives firmly rest. Care should be taken not to have the hives nearer together than 18 inches; the intermediate space will be found very convenient on which to rest the cover, or for supporting an empty hive during the proper performance of any operation.

Mr. Woodbury has his hives arranged on rails, somewhat after the plan before described.

In describing as above the various hives and frames, some hints have been given as to the methods of handling them. This, however, will not suffice for an induction to the mysteries of practical bee-keeping; and we must refer the reader to a subsequent section, wherein the details as to manipulation will be fully explained, and the results of the experience of several distinguished apiarians will be embodied.

## TAYLOR'S IMPROVED COTTAGE HIVE. No. 14.

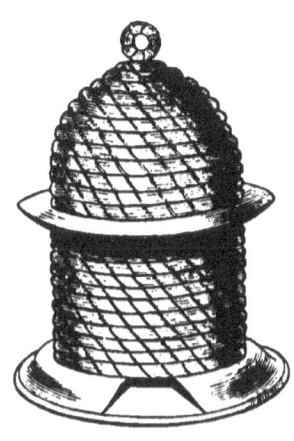

No. 14 hive is similar in principle to No. 8. It is also similar in size, with the exception of being quite straight at the sides. A zinc rim affords protection from the weather. Under the upper straw hive a bell glass is worked. A mahogany adapting board, with a 4 inch hole in the centre corresponding with that of the hive, supplies an even surface for the glass to rest on, and facilitates its removal when full. For particulars for stocking and management see directions for No. 8 and No. 5 hives.

## EIGHT-BAR STRAW HIVE. No. 18.

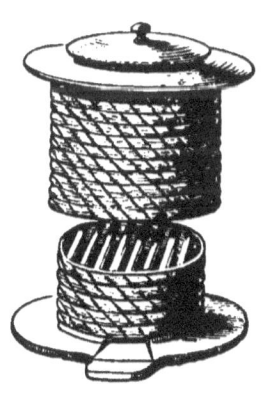

This is an ingenious contrivance of Mr. Taylor's. Hoops are worked in the straw, both at top and bottom of the stock hive, and in the upper hoop are openings cut to receive eight comb bars; each bar is $1\frac{1}{8}$ of an inch wide, with a space of $\frac{1}{8}$ an inch between. Since the introduction of square straw bar and frame hives, these have not been much in request. Considerable inconvenience is found with this hive, because the bars being of unequal lengths they cannot be interchanged one with the other.

The description given of No. 3 hive, and mode of stocking and furnishing it with guide comb, apply to this hive. The large straw hive raised up in the engraving is an outside case; the roof is a large zinc cover. If placed in a bee-house, the outside case and zinc roof are not required.

## NEIGHBOUR'S UNICOMB OBSERVATORY HIVE. No. 20.

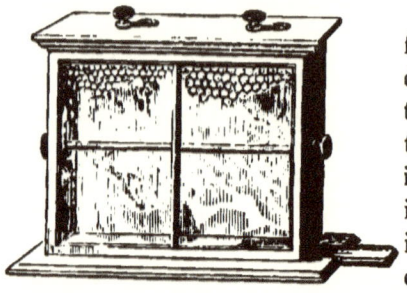

This hive is well adapted for those persons who are desirous of having the opportunity of closely examining the workmanship of these industrious and interesting insects. It is particularly intended for a window recess or an indoor apiary; and will also be found an interesting addition to the green house. An aperture should be cut in the sash corresponding with the entrance to the hive, through which the bees may find egress and ingress, without being able to gain access to the apartment, as on the plan in the case of the Ladies' Observatory Hive.

The Unicomb Hive is constructed of so narrow a width between this plates of glass that it admits of one comb *only* to be built, and at the same time leaves space between the comb and the glass on either side for the bees to pass and repass. It is thus made so that every bee may be exposed to view. The queen forming the most prominent feature of attention, she is readily distinguished by the greater length of her body, as well as by the attention paid to her by the other bees.

The mode of stocking this hive is as follows :—procure a strong swarm if practicable, and let it be first hived into a common straw hive in the usual way. Take care to make the necessary preparations previous to the operation of stocking; for this purpose first unscrew the top and take out the bar intended for the foundation of the comb. A sort of wooden trough, contrived to act as a funnel, is provided with this hive. It is in separate pieces, so as to be packed conveniently, and requires screwing together. The funnel must now be attached to the hive, and care be taken that it is quite firmly fixed—the metal plates are attached for this purpose. Having made these preparations at dusk, dash the bees out on to a cloth or sheet spread on the ground, place four pieces of wood so as to raise the hive a little from the cloth; then invert the hive so as to have the funnel downwards, placing it over the bees, and letting it rest on the four pieces of wood just named. In a short time (say half an hour or an hour) the bees will ascend into the hive; they will go up more quickly if occasionally stirred with the feather of a goose quill.

When the bees have ascended into the hive, take off the funnel, replace the bar, screw on the top board, take it to the place intended for it permanently to occupy, and which has been prepared for it. It will be as well to screen the hive from view for a few days until the bees become settled in their new domicile. Although this hive is con-structed of double glass to keep up a more uniform degree of warmth, still from the cold nature of glass and the close contact into which the bees are brought with it, it is advisable to place flannel between the outer shutters and the glass of the hive on both sides; this is found essential in winter, and very much adds to the comfort of the bees if placed so every cold night during most of the year. In the day time in summer months with the hive being of double glass the whole may be fully exposed to view. If the temperature of the apartment in which the hive stands be kept at 60 degrees, this extra attention will not be so needful. Bees cease to appear disturbed when the exposure to light is continuous. As soon as the bees are settled, comb building will immediately commence, and in about two weeks' time there will be one comb spreading over the whole hive. The queen may be viewed depositing her eggs, and all the usual operations of the rearing of brood, storing of honey, and the building of combs, will be open to full inspection, with perfect ease to the spectator. As an object of lively and permanent interest for the breakfast parlour or conservatory, the " Unicomb Observatory Hive " may be regarded as infinitely superior to an Aquarium or Fernery.

At the Exposition Universelle of 1855 in Paris, we exhibited a hive of this description in full working order. The bees left London on the 5th of July of that year, and were placed in the Exposition on the following morning. An entrance was made for them through the side of the building, as before explained. Our bees had no national antipathies, and they immediately sallied forth to their " fresh fields and pastures new" in the Champs Elysées, the gardens of the Tuileries, the Luxembourg, &c., whence they soon returned laden with luscious store from French flowers.

The Jurors of the Exposition awarded us a Prize Medal for bee-hives.

## WOODBURY UNICOMB HIVE.

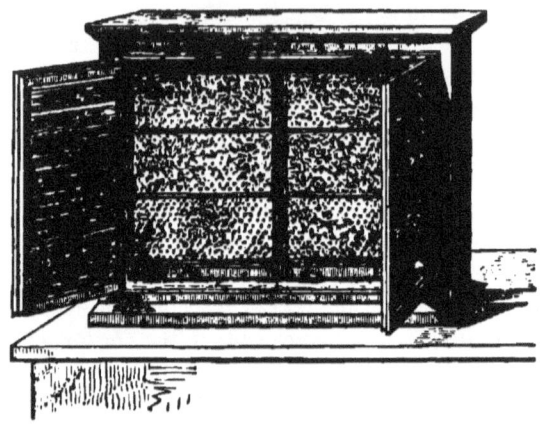

The "Woodbury Unicomb Hive" has many advantages over hives of this kind as previously constructed. The engraving shows the interior compartment divided into six; these are six Woodbury frames. The inner sash opens to admit of hanging up the frames on the notches prepared for them. The width of the hive between the glasses of the sashes is just sufficient to admit of one thickness of comb with space on either side for the bees to pass and repass, the same as in Neighbour's Unicomb. There is, however, a great advantage in the use of this hive; with it, any one possessing a Woodbury box or straw bar and frame hive can readily commence a Unicomb, and as readily put the combs and bees back into the square hive again. The outside shutters on either side are composed of Venetian blinds admitting daylight, but obscuring the rays of the sun. We had the entrance made at one end, as represented in our drawing; this alteration was made after the pattern of the hive from which Mr. Woodbury allowed us to copy. His own was intended to stand wholly out of doors, and had two central entrances, one on either side at the bottom, the hive itself turning on a pivot.

When the hive was being examined on one side, the entrance was closed by a piece of wood inserted in it, and when the other side was brought round to be inspected, the piece of wood withdrawn and placed in the opposite entrance. This was a most ingenious contrivance; but it did not answer our purpose for indoors. When

Mr. Woodbury sent us his hive, we were preparing for the International Exhibition of 1862, and in placing it against the sides of the building, we followed our old plan for ingress by having the entrance at the end. Since that time we have made a considerable improvement by adapting Mr. Woodbury's cleverly contrived turn-table to suit our own hive. Like Mr. Woodbury's hive, ours has two iron wheels, the one fixed to the bottom of the hive, the other fixed to a stout board running the full length of the hive; on these two wheels the whole hive turns. In the centre there is an opening into the hive, with a passage-way running underneath, so that the bees' entrance is in no way affected by the position of the hive, which revolves to suit the convenience of visitors inspecting it. Should the queen with her attendants not be visible on one side, the other side of the comb can be brought into full view, and examined with the same facility as a picture, or as articles are inspected in a shop window. Thus, in the Unicomb Observatory Hive, the sovereign mother, her train of servitors, the drones with their aimless movements, and the crowd of ever busy workers—either building their combs or storing honey—may be always seen as presenting a veritable "Tableau vivant."

Another improvement that we have made upon Mr. Woodbury's pattern is that of accommodating the frames; his was constructed before frame hives were in use, consequently it is only suitable for combs on bars. Our adaptation has necessitated an increase in size. The dimensions are nearly 3 feet square and 7 inches deep from back to front. Provision is made at top for feeding and for working two small flat top glasses for deprivation, which are protected by the weather board.

An alighting board is placed at the centre, close under the entrance, when the hive is located out of doors. When the hive is placed indoors, a passage-way about 18 inches long, covered with glass, is fixed to the entrance, the other end communicating with an opening in the wall or sash; through this the bees find access, an alighting board being fixed outside the building. It is requisite that the passage-way be about this length, in order to allow of the hive turning round clear of the side of the building.

In the summer of 1863 we had ample proof of the success of this hive during its exhibition at the annual show of the Bath and West of England Agricultural Society at Exeter. We selected six combs, and packed them in one of the square box "Woodbury bar and frame" hives, and on arrival at Exeter, Mr. Woodbury assisted us in taking out the frames and placing them in the Unicomb. That being fixed against the boarded side of a shed, we found the covered way a great convenience, and it answered remarkably well; the bees did not seem to be inconvenienced by having to travel through so long a passage. A glass covering admitted a full view of the little labourers as they crowded in, and the sight of them very much enhanced the interest of visitors examining the hive.*

Since the time before mentioned, we have exhibited bees at the meetings of the Bath and West of England Agricultural Society at Bristol, and of the Royal Agricultural Society at Newcastle in 1864. On both occasions further proof was given that this hive admirably answers the purpose intended, and it afforded pleasure and interest to many thousands of visitors.

MODE OF STOCKING.—The Unicomb hive may be stocked in two ways, which have been previously referred to. The bee-keeper may either select the comb upon which the queen is found and put it into the hive, and so form an artificial swarm; or he may take six brood-combs from a hive, and so stock the "Unicomb" at once, which we did for the show at Exeter.

The former plan is, perhaps, the most advisable, because new comb has to be built within the five frames; for, be it remembered, in this case five empty frames must be put in. It is a better plan still if artificial combs are placed in each frame so as to afford an interesting opportunity of watching the formation of the cells therein. The combs are sure to be dark in colour when taken from a stock hive; and new combs being whiter have a better appearance in the hive. The comb upon which the queen was introduced may be taken away

* It may require explanation how it was that we took bees to Exeter, which sounds something like "carrying coals to Newcastle." The reason was this— the garden of our friend, Mr. Woodbury, at Mount Radford, from which we could have been supplied, was so near to the show yard that he was apprehensive a large number of the bees would return to their old hives. Our bees from a distance would, according to their nature, return to their own hive, for bees although they may be moved miles away, take care to mark their new position, and are careful to return to it. Mr. Woodbury lent us a small stock of his Ligurian bees, and between it and our own hive the crowd of visitors divided their attention.

after the artificial swarm has made combs within some of the other five frames; when the queen is on one of the new combs, opportunity may easily be taken for opening the hive and removing the old dark comb. The bees can easily be shaken or brushed off the comb, and will return to the hive. The comb with the unhatched brood may be deposited in any square hive that needs strengthening. We mention this to show how to obtain a hive with entirely fine white comb.

If the possessor of a square Woodbury frame-hive wishes to start a strong Unicomb hive, and does not object to appropriate the stock, he must take out of the Woodbury hive any six combs on the frames, and put the Unicomb in its place so as to receive all the returning bees that happen to be abroad; the remaining four combs, supposing there are ten, may be inserted in any other frame hives in the garden in which are vacancies for them.

We have had this hive in operation, in the manner last described, during the summer of 1863, and found it to answer remarkably well. On a lawn, placed on a suitable ornamental stand, it formed a pleasing object, besides affording great interest and instruction.

In Unicomb hives stocked with a natural swarm (as is generally the plan), there is considerable difficulty in keeping the bees alive through the winter. In a hive where the combs are removeable, no loss of bees need be occasioned. We do not recommend the hive we are now describing as a winter residence for bees. For four months in the year, when bees are most active and when their operations are most interesting, this hive may be brought into use, either of the two plans before described being adopted. An artificial swarm should be put in during May or June, and taken out, in the method before mentioned, and then placed in the square box during the month of September; sometimes it may do for a stock to be put in a month or so earlier, but should never be retained later in this hive. In October we often have cold nights; the bees and brood being in such close contact with the glass, and not able to cluster as is their natural wont, suffer from exposure to the variations of temperature. In some degree to moderate this, we have used treble glass with a space between each square; greater warmth is thus attained, and the view is not intercepted. Opportunity should be taken for cleaning the Unicomb hive when empty, so as to be ready for re-stocking as a new hive in the following

E

summer. The " Unicomb Observatory " hive is one which might have been suggested by the lines of Evans :—

> By this blest art our ravished eyes behold
> The singing masons build their roofs of gold,
> And mingling multitudes perplex the view,
> Yet all in order apt their tasks pursue ;
> Still happier they whose favoured ken hath seen
> Pace slow and silent round, the state's fair queen.

## HUBER'S HIVE. No. 19.

To FRANCIS HUBER—not improperly styled the " Prince of Apiarians,"—we are indebted for more extensive and accurate observations on the habits of the Bee than has been contributed by all other observers since the time of ARISTOTLE.

During the early period of Huber's investigations, he prosecuted them by means of single comb hives which allow of each side of the comb being examined. He found, however, that these had one important defect. The bees could not in these hives cluster together, which is their natural method of withstanding the effects of a reduced temperature. Huber hit upon the ingenious expedient of combining a number of single comb frames so as to form one complete hive, which could be opened in order to expose any particular comb, without disturbing the rest. From the manner of the opening and closing of this hive, it has generally been called the " Leaf or Book Hive." The division separating each comb is joined both back and front with " butt hinges," fastened with a moveable pin, on withdrawing which, at either side, each comb and the bees on it may be inspected as easily as if in a single comb hive. Huber's Leaf Hive is thus in appearance as if several of those " History of England " back-gammon chess boards were set up on end together. The floorboard on which the hive stands is larger than the hive when closed, so as to allow of its being opened freely at any particular " volume." An entrance way for the bees is hollowed out of the floorboard as in other hives. There is a glass window in each end of the hive, which is provided with a shutter.

There is, however, one serious objection to Huber's hive, which, though not noticed by him or his careful assistant, has prevented its general use. That is, the difficulty there is in closing it without crushing some of the bees, a catastrophe which, by exasperating their comrades, is certain to interfere with any experiments. There is no such risk in the Bar and Frame Hive, whilst in it every facility possessed by Huber's is retained; so that we strongly recommend scientific apiarians, in preference to Huber's, that they should use some kind of Bar and Frame Hive. We have here introduced a description of Huber's Leaf Hive (and should be glad to exhibit one) for the sake of its historic interest in connection with apiarian science. The invention was invaluable for Huber himself, and it suggested to other apiarians the adoption of the present plan of vertical bars and frames.

The character of Huber, and the circumstances under which he pursued his observations, are so remarkable, that we need scarcely apologize for stating a few particulars respecting him here. He was born at Geneva in July, 1750, his family being in honourable station and noted for talent. Just as he attained to manhood he lost his sight, and remained blind to the end of his days. This apparently insuperable obstacle in the way of scientific observation, was overcome by the remarkable fidelity with which Burnens, his assistant, watched the bees, and reported their movements to Huber. Madame Huber also who, although betrothed to him before his calamity, had remained constant in her affection, assisted in the investigations with great assiduity during their long and happy wedded life. We quote the following from "Memoirs of Huber," by Professor de Candolle :—

" We have seen the blind shine as poets, and distinguish them-selves as philosophers, musicians, and calculators; but it was reserved for Huber to give a lustre to his class in the sciences of observation, and on objects so minute that the most clear-sighted observer can scarcely perceive them. The reading of the works of Reaumur and Bonnet, and the conversation of the latter, directed his curiosity to the history of bees. His habitual residence in the country inspired him with the desire, first of verifying some facts, then of filling some blanks in their history; but this kind of observation required not only the use of such an instrument as the optician must furnish, but an intelligent assistant, who alone could adjust it to its use. He had then a servant named Francis Burnens, remarkable for his sagacity and for the devotion he bore for his master. Huber practised him in the art of observation, directed him to his researches by questions adroitly combined, and aided by the recollections of his youth, and by the testimonies of his wife and friends, he rectified the assertions of his assistant, and became enabled to form in his own mind a true and perfect image of the manifest facts. 'I am much more certain,' said he, smiling, to a scientific friend, 'of what I state than you are, for you publish what your own eyes only have seen, while I take the mean among many witnesses.' This is, doubtless, very plausible reasoning, but very few persons will by it be rendered distrustful of their own eyesight."

The results of Huber's observations were published in 1792, in the form of letters to Ch. Bonnet, under the title of "Nouvelles Observation sur les Abeilles." This work made a strong impression upon many naturalists, not only because of the novelty of the facts stated, and the excellent inductive reasoning employed, but also on account of the rigorous accuracy of the observations recorded, when it was considered with what an extraordinary difficulty the author had to struggle.

Huber retained the clear faculties of his observant mind until his death, which took place on the 22nd of December, 1831. Most of the facts relating to the impregnation of the queen, the formation of cells, and the whole economy of the bee-community as discovered and described by Huber, have received full confirmation from the investigations of succeeding naturalists.

## III. EXTERIOR ARRANGEMENTS AND APPARATUS.

BEE-HOUSE TO CONTAIN TWO HIVES. No. 39.

Front View of Bee-House.

There is no contrivance for protecting hives from the weather so complete as a bee-house one, which also admits of an easy inspection of the hives ranged therein. This arrangement is especially convenient for lady bee-keepers.

The folding doors behind the bee-house have only to be opened, and the hives are at once exposed to full view. If the cover of the bee-house be lifted as well as the shutters opened, the hives and the glasses may be deliberately inspected, without any danger of molestation from the bees. Thus the progress made by the busy multitude in building and filling their combs may be watched by the bee-keeper, from day to day with great and increasing interest.

Back View of the Interior.

Here our engraving shows the back view of the bee-house, the interior being furnished with two of our No. 5 cottage hives. Two suspended weights will be noticed; these are to balance the top hives which cover the glasses; the cord for each, runs on pulleys, so that the covers can be easily raised and as easily shut down again when the inspection is finished. We may here remark, that it is not well to keep the glasses long exposed to full light and view.

The front of the bee-house being closely boarded, a passage way is contrived for the bees by which they have egress and ingress, without being able to gain access to the house. The hives must be kept close to the front boarding of the house, so as to prevent the opening of any crevices which the bees might mistake for the entrance to their hives, and so find their way into the house. The front view of this bee-house shows the ordinary contrivance for entrance ; the sliding zinc entrances may also be advantageously fixed as shown in the engraving of a " Bee-house to contain twelve hives." In many parts of the country, hives and honey are some-times stolen from the garden ; the bee-houses we furnish have a lock and key to prevent depredations of this kind.

Care must be taken to keep the bee-houses free from spiders and other insects. In some districts ants are numerous and trouble-some. The plan we recommend for excluding them is to put some pitch round the four supports of the bee-house ; or, better still, strips of loose flannel or other woollen material that is absorbent, which have previously been soaked in lamp-oil. We use sperm oil, being the slowest drying oil we know of. A piece of string will keep the flannel close to the wood, and then neither ant or any other insect will pass up ; so that by this simple means the hives may, so to speak, be insulated and placed beyond their reach. As the oil dries up it can easily be renewed. We have found this an effectual remedy against these insidious enemies of bees.

## BEE-HOUSE TO CONTAIN TWELVE HIVES. No. 40.

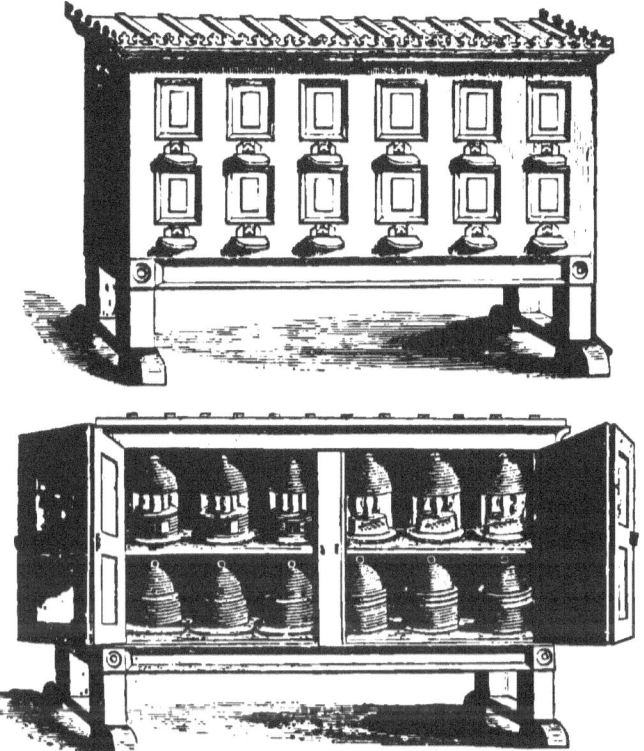

Back View of the above, shewing the Interior.

Where economy of room is a consideration, we fit up bee-houses with a double row of hives, one above the other. Our engravings show the back and front of a house of this kind, having an ornamental zinc gutter to prevent the wet from dripping on to the alighting board.

When a number of hives are thus together, we colour the alighting boards differently, so that bees may have a distinctive mark by which each may know its own home, and not wander into its neighbour's house. Bees readily enough receive a honey laden labourer into a hive; but if the wanderer be poor and empty, he will be promptly repulsed, and may have to forfeit his life for his mistake. Queens returning from their wedding trip, are liable to mistake their hive if all the entrances are so much alike that a noticeable difference is not easily apparent. A queen entering a hive already supplied with a fruitful sovereign would be certain to be killed. The loss to the hive to which the queen belonged is a most serious one. Hives are often made queenless from this cause, and thereby reduced to utter ruin; the bee-master perhaps attributing his loss to something altogether different.

## BEE-HOUSE TO CONTAIN NINE HIVES. No. 41.

This engraving represents a bee-house adapted for having a number of hives in a limited space. Three rows of hives, one above the other.

We do not recommend a bee-house of this construction; it is difficult to erect one to afford space for super hives, without it being so inconveniently high as to be liable to be blown over by strong winds.

Hives thus located in a bee-house are not exposed to so much change of temperature and the stocks generally pass the winter well.

Here we may introduce the meditations of a German apiarian, as he describes the advantages of a bee-house for the bees, and his own pleasure in watching over his pets in the winter, as they dwell so comfortably and safely. It is true that HERR BRAUN associates still choicer delights with the simple pleasures of bee-keeping, but as MR. WOODBURY has not excluded the higher theme from his translation, we need not hesitate to quote the whole:—

### EVENING THOUGHTS IN JANUARY.

*(Translated from the German of* ADALBERT BRAUN.)

BY "A DEVONSHIRE BEE-KEEPER."

| | |
|---|---|
| Within my little garden<br>  Stands also a bee-house,<br>And bees therein protected<br>  From sly tomtit or mouse. | How quietly they're sitting!<br>  And little trouble give,<br>Beyond the needful watching,<br>  How undisturb'd they live; |

That all, indeed, are living
  In strong unbroken health,
And, in the brood-nest hanging,
  Consume their hoarded wealth—

That in the dwindling store-room
  Sufficient stores remain,
Until the rape-plant donneth
  Its blossom dress again.

Thus daily do I visit
  My garden and my bees,
Neglecting thereby often
  My dinner and my ease.

Thank God! they all were humming
  Within their hives to-day;
Nor could I find a symptom
  Of hunger or decay.

And yet what ardent longing
  I feel, O Spring, for thee!
My darlings' gleesome frolics
  Are happiness to me!

How would this anxious longing
  Consume my very breast,
But for a little being
  So full of love and jest,

In heat or cold who prattles
  Around me ev'ry day,
And stills the throes of longing
  By commune blithe and gay.

Ye bee-keepers can value
  A joy that is complete;
It is my wife—the darling—
  Whose lips are honey-sweet.

With e'en the richest bee-stand
  Were joy and pleasure gone,
If my heart's queen were wanting
  And I left here alone.

Thus her I love and honour,
  No difference have we,
But oft-times go together,
  Our little pets to see.

Her kisses sweet removing
  All sorrow from my breast,
And honied joys surrounding
  Proclaim us highly blest.

T. W. WOODBURY, *Mount Radford, Exeter.*

EXTERIOR OF AN APIARY.

As originally erected in the Zoological Gardens, Regents Park.

INTERIOR OF THE ABOVE.

May be taken as suggestive for the construction or appropriation of rooms
for larger Apiaries in summer houses or other outbuildings.

## ZINC COVER. No. 37.

This is a simple and inexpensive covering for a No. 5, or other cottage straw hive when exposed in the garden. It fits close on to the upper hive, coming sufficiently low to protect from the sun and rain, without obscuring the whole hive.

These covers are painted green, that colour being generally preferred.

## ORNAMENTAL ZINC COVER. No. 38.

The annexed engraving of the Ornamental Zinc Cover renders but little description necessary. The illustration shows one of our No. 5 improved cottage hives on a stand. Three clumps of wood must be driven into the ground, and the three iron rods supporting the covering made fast to them with screws. There are screw holes in the feet of the iron rods for the purpose ; when thus secured, but little fear need be entertained of its being blown over by high winds.

In the roof two pulleys are fixed, so that by attaching a cord, the upper hive covering the bell glass supers may be raised with facility for the purpose of observing the progress made by the bees.

The Ornamental Zinc Cover will form a pleasing object in the flower garden when placed in a suitable position on the grass plot. It is painted green ; the iron rods are of such a length as to support the roof at a convenient height from the ground.

## COVER OF ZINC. No. 29.

This zinc cover introduced by H. Taylor, Esq., for his cottage hive (No. 14) will also be found useful as a protection from wet for many other descriptions of round straw hives.

## BELL GLASSES.

| 25. | 26. | 27. |

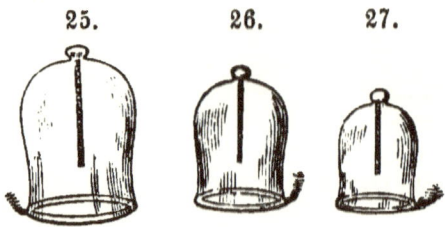

25.  To contain 10 lbs., 10 inches high, 7 inches wide.
26.  To contain 6 lbs., 7 inches high, 5½ inches wide.
27.  To contain 3 lbs., 5 inches high, 4 inches wide.

These bell glasses are used in the hives before described. No. 25 is for Nutt's Hive (No. 1); No. 26 is for our Improved Cottage Hive (No. 5); No. 27 is a very small glass, one that is not often used, and which we do not recommend. Bees will generally fill a middle sized glass quite as soon as one so small as this.

## BELL GLASSES. No. 24.

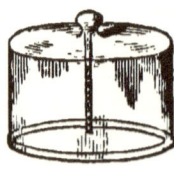 

These glasses have been introduced by Mr. Taylor, and are recommended as preferable to deep narrow glasses.

The drawings will show that they are straight at the sides, flat at the top inside, with a knob outside to take hold by, through which is an ½ inch opening to admit a ventilating tube.

The larger is 6 inches deep, 12 inches wide; smaller 5 inches deep, 9½ inches wide.

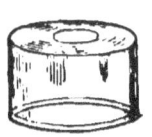

The late Mr. J. II. Payne, of Bury, author of the "Bee-keeper's Guide," introduced another glass. It has a 3 inch hole in the centre, the purpose of which is to tempt bees to produce additional and larger stores of honey. It is to be used as follows :—when a bell glass is half or quite filled, raise it, and place the Payne's glass over the hole of the stock hive, with the filled glass on it over the 3 inch hole. The bees will bring their combs through, and thus Mr. Payne found that they would store more honey than if the bell glass were removed and another empty one put in its place. Of course the first glass must be smaller in diameter than the Payne's glass, so as to rest upon it.

## BELL GLASS. No. 28.

This is a glass super to be placed on the hive in a similar way to the bell glasses already alluded to. It has the advantages of being straight at the sides, flat at top, and without a knob; so that when filled it may be brought on to the breakfast table, inverted, on a plate. The glass lid shown in the engraving forms a cover, and fits over outside, so as not to interfere with the combs within. These flat top glasses, like those with a knob, have a hole through which a zinc ventilating tube is inserted.

## GUIDE COMB FOR GLASSES.

In some of our previous allusions to the best mode of inducing bees to commence working in glasses, we have recommended attaching guide comb. We will now more particularly explain how this attraction can be best applied. We have already shown how bees may be induced to make use of guide combs fixed to bars, and the same principle is applicable to glasses. These may be filled with great regularity by adopting the following directions, which we believe have never before appeared in print :—

Procure a piece of clean new empty worker honeycomb which has not had honey in it (because honey will prevent adhesion to the glass); cut it up into pieces of about three quarters of an inch square. Gently warm the exterior of the glass; this we find is best done by holding the glass for a short time over the flame of a candle; then apply one of the pieces of empty comb inside at the part warmed, taking care in fixing it that the pitch or inclination of the cells is upwards—in fact, place the guide comb in the same relative position that it occupied in the hive or glass from which it was taken. There is some danger of making the glass too warm, which will cause the comb to melt, and the wax to run down the side, leaving an unsightly appearance on the glass; this should be carefully avoided, and a little experience will soon enable the operator to determine the degree of warmth sufficient to make the comb adhere without any of it being melted. It is hardly necessary to state that only the very whitest combs ought to be used. A short time should be allowed before changing the position of the glass, so that it may cool sufficiently to hold the comb in its place. Six or eight pieces may thus be fixed, so that  when the glass is filled, it will present a star shape, all the combs radiating from the centre. The annexed engraving shows the appearance of a glass as worked by the bees, in which guide combs were fixed in the manner described above. The drawing was taken from a glass of our own filled after being thus furnished.

In the Old Museum at the Royal Gardens, Kew, may be seen a Taylor's glass presented by us, some of the combs in which are elongated on the outside to the breadth of six inches. We believe that not only does a glass present a much handsomer appearance when thus worked—and will, on that account, most fully reward the trouble of fixing guide comb—but that more honey is stored in the same space and in less time than if the glass be placed on the hive merely in a naked condition for the bees to follow their own devices.

This mode of fixing guide comb does not solely apply to this shaped glass, but is equally useful for all kinds of glasses. It is introduced in connection with No. 28, because that glass having a flat top and no knob, the regularity is more clearly apparent.

The working of bees in the bell glasses illustrates how tractable their disposition really is if only scope be allowed for the due exercise of their natural instinct. They have no secrets in their economy, and they do not shrink from our constant observation as they daily pursue their simple policy of continuous thrift and persevering accumulation. Yet it is only owing to the labours of successive inventors that we are now enabled to watch "the very pulse of the machine" of the bee-commonwealth :—

> "Long from the eye of man and face of day,
> Involved in darkness all their customs lay,
> Until a sage well versed in Nature's lore,
> A genius formed all science to explore ;
> Hives well contrived, in crystal frames disposed,
> And there the busy citizens disclosed."—MURPHY's *Vaniere.*

## THE NEW BOTTLE FEEDER. No. 44.

It has long been acknowledged that the best mode of feeding bees is through an opening at the top of the stock hive. The "new bottle feeder" is a simple and good means of administering food when a stock requires help in that way. Any kind of hive that has an opening at the top may thus be fed ; bees can take the food from it without leaving the hive. Another important feature is the cleanliness with which liberal feeding can be accomplished ; and few operations require more care than does feeding. If liquid sweet is left hanging about the hive, it tempts robber bees, and when once the bees of an apiary have had a taste, there is no knowing where their depredations will stop ; they resolutely attack and endeavour to rob other hives, fighting and killing one another to a considerable extent. Even if no hives be completely destroyed, weakness from loss of numbers will be the portion of most, if not all, the hives in the garden.

The morals of our favorites are here a good deal at fault, for the strongest hives, when their inordinate passion is thus stirred up by the carelessness or want of knowledge of the bee-keeper, attack and prey upon the weak. To be "forewarned is to be forearmed "— and "prevention is better than cure." We strongly recommend closely covering up the feeder ; one of the middle size bell glasses put over it makes a close fitting cover, should the regular cover to the hive not be sufficiently tight : when bees are not kept in a bee-house—and are on that account more accessible—this extra care is more particularly needed. The right time for feeding is in the autumn or spring. A stock of bees at Michaelmas ought to weigh 20 lbs. exclusive of the hive, and if then it weigh less, the deficiency should be made up by artificial food. It is not wise to defer doing this until later in the season, because it is important that when the food is placed in the cells, the bees should seal it up, and a tolerably warm temperature is required to enable them to secrete the wax for the delicately formed lids of the cells. If the food remain unsealed, there is danger of its turning sour and thereby causing disease among the bees. It is not well to feed in mid-winter or when the weather is very cold. Bees at such times consume but little food, being in a state of torpor, from which it is better not to arouse them.

A little food in the spring stimulates the queen to lay more abundantly, for bees are provident, and do not rear the young so rapidly when the supplies are short. In this particular the intelligence of bees is very striking; they have needed no Malthus to teach them that the means of subsistence must regulate the increase of a prosperous population:

> "The prescient female rears the tender brood
> In strict proportion to the hoarded food."—EVANS.

Judgment has, however, to be exercised by the apiarian in giving food, for it is quite possible to do *mischief by over feeding*. The bees when over-fed will fill so many of the combs with honey that the queen in the early spring cannot find empty cells in which to deposit her eggs, and by this means the progress of the hive is much retarded, a result that should be guarded against.

The following directions will show how the bottle feeder is to be used:—Fill the bottle with liquid food, place the net fixed on with an India-rubber band over the mouth, place the block over the hole of the stock hive, invert the bottle, the neck resting within the hole in the block; the bees will put their proboscises through the perforations and imbibe the food, thus causing the bottle to act on the principle of a fountain. The bottle being glass, it is easy to see when the food is consumed. The piece of perforated zinc is for the purpose of preventing the bees from clinging to the net, or escaping from the hive when the bottle is taken away for the purpose of refilling. A very good syrup for bees may be made by boiling 6 lbs. of honey with 2 lbs. of water for a few minutes; or loaf sugar, in the proportion of 3 lbs. to 2 lbs. of water, answers very well when honey is not to be obtained.

## ROUND BEE FEEDER. No. 10.

Round bee-feeders are made of zinc and earthenware; 8 inches across, 3 inches deep. The projection outside is a receptacle for pouring in the food; the bees gain access to the feeder through a round hole, which is placed either at the centre or nearer one side, whichever may best suit the openings on the top of the stock hive. The feeder occupies a similar position to that of the glasses or cap hives in the gathering season. A circular piece of glass, cut so as to fit into a groove, prevents the bees escaping and retains the warmth within the hive, whilst it affords opportunity for inspecting the bees whilst feeding.

The feeders were originally only made of zinc; but some bee-keepers advised the use of earthenware, and a few have been made to meet the wishes of those who give the preference to that material.

When the bees are fed from above in this manner, the feeder is kept at a warm temperature by the heat of the hive. In common hives cottagers feed the bees by pushing under the hive thin slips of wood scooped out, into which the food is poured. This plan of feeding can only be had recourse to at night, and the pieces of wood must be removed in the morning. By feeding at the top of the stock hive any interruption of the bees is avoided. For further instructions on this head see the directions given for using the bottle feeder.

## A ZINC FOUNTAIN BEE FEEDER. No. 15.

We invented the fountain bee-feeder so that a larger supply of liquid food might be given to a hive than is practicable with the No. 10 round feeder.

The liquid honey is poured in at the opening, which unscrews; whilst being filled, the inside slide closing the opening through which the food passes into the feeding pan, should be shut down. When the reservoir is filled, the screw is made fast, and the slide being withdrawn, a wooden float pierced with small holes, through which the bees take the food, forms a false bottom, and rises and falls with the liquid.

This feeder being on the syphon principle, like a poultry or bird water-fountain, is supplied from the reservoir until that is empty. A piece of glass is fixed in the side of the reservoir, in order that the bee-keeper may see when it is emptied. A flat piece of glass on the top prevents the bees from escaping, and through it they may be inspected whilst feeding. The bees find access to the feeder on to the perforated float through the central round hole, which is placed over a corresponding hole in the stock hive.

### NUTT'S DRAWER FEEDER.  No. 9.

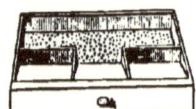 This is the feeding drawer, alluded to in the explanation of the management of the No. 1 Nutt's Collateral Hive, for feeding at the bottom of the middle or stock box. The feeder is made of tin, and of a square form, so as to fit the drawer (see page 19).

### HONEY CUTTERS.  No. 13.

 Honey Cutters are used for removing comb from boxes and glasses without damaging it. The flat bladed knife is for disconnecting the combs from the sides; the hook shaped one is for the same purpose, to be applied to the top or horizontal part of the box or glass.

### FUMIGATOR.  No. 11.

 This Fumigator is a tin box, somewhat like a pepper box upon a foot. It is a simple adaptation of the fumigating apparatus described by Mr. Nutt, and is used in the following manner. Have a straw hive or other vessel ready, that will match in circumference the hive intended to be fumigated. If the empty hive have a conical top, it will not remain crown downwards without a rest; in this case, it will be convenient to invert it on a pail. Having ascertained that the hive to be operated upon and the empty one in its reversed position nearly match in size, take half a packet of the prepared fungus, fire it well, and place it in the box or fumigator; placing this in the

centre of the empty hive, then bring the occupied hive directly over, so as to receive the fumes of smoke. To keep all close, put a wet cloth round the place where the two hives meet. In a minute or two the bees may be heard dropping heavily into the lower empty hive, where they lie stupefied. After a little while, the old hive may be tapped upon to make the bees fall more quickly. On removing the upper hive, the bees from it will be found lying quiet at the bottom of the lower one. Place a sheet on the ground, and spread the bees on it, then with a feather sort them over, in order to pick out the queen-bee. As soon as the queen is found, then pour the rest of the lethargic swarm from off the sheet back into the inverted hive again. The stupefied bees must now be sprinkled freely with a syrup made of honey and water, or sugar and ale boiled together. Some apiarians recommend a few drops of peppermint to be mixed with the syrup, in order to drown the peculiar odour which is special to each hive of bees,—this is more necessary when two hives of bees are fumigated and whilst under the influence of smoke are well mixed together. The hive containing the bees with which it is intended to unite the stupefied bees, must now be placed on the top of that containing the latter, just as the hive was from which they have come. A wet cloth must be fastened round the two hives, so as to prevent any of the bees from escaping. The hives in this position must be placed where they are not likely to be knocked down or meddled with. The fresh bees in the upper hive, attracted by the scent of the bees besmeared with honey, go down and commence to lick off the sweets from the sprinkled sleepy ones. The latter gradually revive, when all get mingled together, and ascend in company to the upper hive, where they live as if they had not been separate families.

The two hives should be left undisturbed for twenty-four hours, then the upper hive may be removed and placed immediately on the spot from whence it was brought. The reason the queen is recommended to be taken is to prevent any fighting.

The queen should be kept alive and fed as long as she will live, in case any harm should befall the sovereign of the other community.

## THE FUMIGATOR.   No. 12.

The Fumigator is useful for several purposes. When a frame hive has to be disturbed it is requisite to raise the lid and blow a little smoke into the hive, so as to check the angry passions of the bees. If it be desirable to stupefy the bees, ignited fungus must be placed in the box, and the flattened end applied to the entrance of the hive; the smoke is then blown in—either with bellows or by applying the mouth of the operator,—taking care to close all openings through which the smoke can escape. The bees fall down stupefied, generally in about ten minutes, but the effect varies according to the populousness of the hive and the quantity of comb in it. The projected operations must now be performed speedily, as activity will soon be regained. See preceding directions.

## THE BEE DRESS OR PROTECTOR.   No. 31.

All operations connected with the removal or the hiving of bees should be conducted with calmness and circumspection. Bees, although the busiest of creatures, have great dislike to fussiness in their masters, and become irritable at once if the apiarian lets them see that he is in a hurry. Hence, there is great advantage in having the face and hands covered whilst at work amongst the bees; for when the operator knows he cannot possibly be stung, he can then open his hives, take out the combs, gather in his swarms, or take the honey, with all the deliberation of a philosopher. Various kinds of bee dresses have been contrived; one that we keep ready in stock is of very simple construction. It is made of strong *black net*, in shape like an inverted bag, large enough to allow of a gentleman's wide-awake or a lady's hat being worn underneath. The projection of the hat or cap causes the dress to stand off from the face; and the meshes of the net, though much too small for a bee to penetrate, are wide enough to allow of clear vision for the operator. An elastic band secures the dress round the waist; the sleeves also —made of durable black calico—

are secured at the wrists by a similar method. The hands of the bee-master may be effectually protected with a pair of India-rubber gloves, which should be put on before the dress is fastened round the wrists. This kind of glove is regularly used by photographers, and allows of perfect ease in manipulation.

Thus a very simple and inexpensive means of protection will enable even a novice in bee-keeping to make his observations and conduct his experiments under a sense of perfect security. Still he need not be careless as to the feelings of his bees ; his success and their comfort will be promoted by his " handling them gently, and as if he loved them." " Familiarity" between bees and their master " breed" not " contempt," but affection.

Any sudden or clumsy movement which jars the combs or frames will excite the bees, and if but one should be crushed, the odour of their slaughtered comrade rouses them to a pitch of exasperation. Their powers of smell are very acute. The best time for most operations is in the middle of a fine day.

### ENGRAVED PRESSING ROLLER. No. 46.

#### FOR THE GUIDANCE OF BEES IN THE CONSTRUCTION OF HONEY-COMB ON THE BARS.

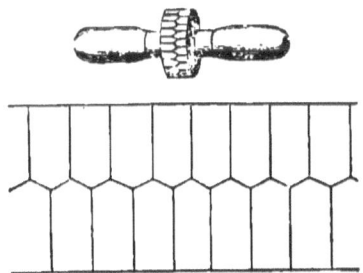

This is an engraved metal roller, which when applied to the coated underside of a comb-bar leaves an impression like the diagram shown. The wax having been spread on the flat bar, the roller, heated by being put into hot water, is then heavily pressed over it. The roller has two wood handles, so that considerable pressure may be given to it. The roller is a little less than 2 inches in diameter, ¼ of an inch wide, and the length from handle to handle is 6 inches. The diagram shows the full size of the impressions as left on the wax after passing the roller along the comb bar in the manner above described. It is a contrivance invented in Switzerland, and exhibited in the International Exhibition of 1862, when the pattern roller was purchased by ourselves.

The bars of a hive prepared with these markings in wax afford ready made foundations for regular combs, which very much facilitate the operations of the bees.

## IMPRESSED WAX SHEETS FOR ARTIFICIAL COMB.

These artificial partition walls for combs are sheets of genuine wax, about the substance of thin cardboard. They receive rhomboidal impressions by being pressed between two metal plates, carefully and mathematically prepared and cast so that the impressions are exactly the same size as the base of the cells of a honey-comb. An inspection of a piece of comb will show that the division of the opposite cells is made by a thin partition wall common to both ; now the substance of this is said to be only the one hundred and eightieth part of an inch, whilst the artificial ones we are recommending are between the thirtieth and fortieth part of an inch, more than four times the thickness of the handy-work of the bees themselves. It would, indeed, be vain to attempt to furnish sheets of wax at all approaching their own delicate fabric ; the impressed sheets are quite as thin as they can be to bear the handling which is requisite for fixing them in the hives. We find, however, that the thickness is no disadvantage ; the bees speedily excavate and *pare the artificial sheet* so as to suit their *own* notions of the substance required ; then, with admirable economy, they use the surplus thus obtained for the construction of the cells. After a sheet has been partly worked at by the bees, it is interesting to hold it up to the light and observe the beautiful transparency of that part of it contrasted with the opaqueness of the part not yet laboured upon.

When it is considered, as writers tell us, that more than fourteen pounds of honey are required for the secretion and elaboration of a single pound of comb, it will not be difficult to form a just estimate of the value of this invention, which thus furnishes cheap and excellent assistance to our industrious favorites. It also shows the bee-keeper that all clean empty combs should be carefully preserved, and considered as valuable stock. Another great advantage that it affords us is, that it renders us independent of *guide comb*, which is not always obtainable. When a sheet or a strip of this impressed wax is properly fixed to the comb bar, it is *certain* to be the guide and foundation of a straight comb. This invention has been derived from Germany, where it has been adopted many years with success. At the International Exhibition of 1862, we purchased the metal plates or castings, so as to manufacture the impressed sheets with which we are now able to supply our customers; and, after the careful trials we have made, we have great confidence in recommending them.

In the season of 1863 we furnished a Woodbury glass super with the wax sheets fixed to the bars, in the manner hereafter to be explained, and it was truly astonishing to see the rapidity with which these sheets of wax were worked into comb. Receptacles were quickly made ready for the storing of honey, and the new combs soon became beautifully white; for although the artificial wax has a yellow tinge, yet after being worked at and made thinner, it is as good in colour as ordinary combs. For supers we cut the wax plates in half, making one serve for two bars.

We have received from Germany the following directions for the fastening of the artificial plates to the comb bars. Hereafter will be described a plan which we have adopted, and to which preference is given.

(Translation) "The unstamped edge of the plate receives incisions half an inch distant from one another made with a sharp knife, the plate having been a little warmed. Then it is pinched between two equally strong ledges which have been well moistened; the projecting edge of the plate which received the incisions is alternately bent to the right and to the left. The comb bars are well besmeared with artificial sticking wax (a mixture of two parts of wax and one part of American resin), and is well warmed at a fire. Afterwards the besmeared side is laid upon the bent end of the plate, and pressed to it as firmly as possible. A small wooden ledge, besmeared with sticking wax and fastened by means of pressure to the lower edge of the plate, prevents it from bending, which sometimes happens when the bees work it."

To carry out the directions here given, it is necessary to warm the besmeared comb-bar at a fire; the wax plate has also to be warmed. Having tried this plan, and found inconvenience attending it, especially from the wax curling with the heat and the difficulty

of making it stick firm,—to say nothing of the uncomfortableness
of performing the operation before a fire on a hot day in July,—we
began to consider if a little carpentering might not do the work
better and more pleasantly, and adopted the following plan :—
We split or cut the comb bars of the Woodbury super in
half lengthways, and taking the unstamped edge between the two
strips, join them together again by small screws at the side,
confining the wax plate tightly in the centre, with no possibility of
its falling down. Where frames are used, of course, the bar could
not be cut in two (except with the " compound bar and frame,"
where the bar being loose, it might be as easily managed). The
plan we adopt with an ordinary frame is to saw out an opening
about an inch, or an inch and a half from either end where the
sides are morticed in; this opening we make with a keyhole saw.
Through it the wax plate is easily put, and with a heated iron
passed over the upper side of the bar, is made sufficiently firm.
If the wax plates are too large, a portion may easily be cut off; an
opening of full 11 inches long can be made without materially
weakening the bar and frame.

The wax plates must not extend to the bottom of the frame ; a
space of at least one inch should be left for expansion, because the
bees in working the plate stretch it down lower. We also use a few
pins firmly pressed into the frames, and long enough to reach the
edge of the plate, for by fixing three or four pins on either side,
both at the sides and at the bottom, the plate may be held in an
exactly central position within the frame. As before mentioned,
when these directions are carried out, there is no fear of being
troubled with crooked combs on bars.

The secretion of wax, and the method of its adaptation by the
bees, is thus admirably described by Evans :—

> Thus filtered through yon flutterer's folded mail
> Clings the cooled wax, and hardens to a scale.
> Swift at the well-known call, the ready train
> (For not a buzz boon Nature breathes in vain,)
> Spring to each falling flake, and bear along
> Their glossy burdens to the builder throng.
> These, with sharp sickle, or with sharper tooth,
> Pare each excrescence, and each angle smooth,
> Till now, in finish'd pride, two radiant rows
> Of snow-white cells *one mutual base* disclose;

Six shining panels gird each polish'd round,
The door's fine rim, with waxen fillet bound,
*While walls so thin, with sister walls combined,*
*Weak in themselves, a sure dependance find.*

\* \* \* \* \* \* \*

Others in firm phalanx ply their twinkling feet,
Stretch out the ductile mass, and form the street,
With many a cross-way, path, and postern gate,
That shorten to their range the spreading state.

## IV. MANIPULATION AND USES OF BAR AND FRAME HIVES.

Having at page 36, given a description of the mechanical arrangements of bar and frame hives, the next thing is to describe the mode of introducing the bees, and of thus bringing the humane and scientific hives into operation. The swarm should be first hived into a common straw hive from the bough or shrub upon which they may have alighted; place this hive, —into which we will suppose the bees have been shaken,—on the ground, propped up on one side with a brick or a flower pot, or anything of the sort that may be handy, in order that straggler bees may join the swarm. The spot selected for this should be as shady an one as can be found, near to the place where the swarm settled; or it may be shaded from the rays of the sun by fixing matting on two poles so as to prevent the heat falling on the hive; spread a sheet or cloth on the ground where an even surface can be obtained; stake this sheet down at the four corners, to prevent ruts and inequalities, which are great hindrances to the bees going into the bar and frame hive; place the latter upon the sheet without its floorboard, having its front raised on blocks or sticks rather more than an inch,—not more, otherwise the bees will cluster and attach themselves to the lower part of the frames instead of going up between. These preparations will perhaps occupy ten minutes, by which time the swarm will have become settled and tolerably quiet. Then with a sharp rap precipitate the bees out of the straw hive on to the sheet immediately in front of the frame hive; give the straw hive another knock so as to dislodge all the bees, and then take it quite away, otherwise they may, if it be left near, perversely choose to go into that instead of the one desired. In some cases, as when the swarm has to be brought from a distance, and procured from a cottager about whose skill in carrying out these directions there may be misgivings, it is best to give instructions that the swarm be brought home after sunset, and then the foregoing directions for inducing the bees to tenant the frame-hive may be better carried out; for ourselves, we much prefer the evening for the purpose. A little water sprinkled over them from a watering pot is likely to induce the bees to quit the ground and go up into the hive more quickly.

Mr. Langstroth, in his admirable book "The Hive and Honey Bee," writes "If they are too dilatory in entering the new hive, they may be gently separated with a spoon or leafy twig, where they gather in bunches on the sheet, or they may be carefully 'spooned up' and shaken out close to the the front of the hive. As these go in with fanning wings, they will raise a peculiar note, which communicates to their companions that they have found a home, and in a short time the whole swarm will enter, without injury to a single bee." In the *Journal of Horticulture*, Mr. Woodbury says "If combs be fixed in the frames, the crown board may be removed, and the cluster knocked out of the straw hive on to the top of the exposed frames. The bees will disappear between them with the utmost alacrity, delighted to have met with a ready furnished dwelling, and the top or crown board having been replaced, the hive should at once be removed to the position it is intended permanently to occupy."

No one should attempt these operations without being protected by a *bee dress*, and a pair of India-rubber gloves,—such as those used by photograpers,—which are sting proof. Some persons also take the precaution of tieing strings round the ancles of the trowsers, lest some straggler should determine to attack the outposts of the enemy, which, to say the least, might perplex the operator in the midst of his task. Elastic India-rubber bands are good for this purpose, or a pair of "knickerbockers" would be useful. If Wellington boots are worn, the trowsers may be tucked within the leather, in which case no bee can molest the operator, and no string or band will be needed. Practice makes perfect in bee tending as in other matters, and when a light hand is gained, there is little danger of the apiarian being stung.

If the weather be wet the next day or so after hiving, it will be well to give a little assistance to the new colony in the shape of food, for although when a swarm leaves a hive almost every bee composing it fills itself with honey, we have known not a few instances in case of very wet weather, in which the whole swarm has been starved for the want of this little timely help. Of course, the first work of the bees is to build themselves combs, and these combs being produced by the secretion of wax from honey, a great drain upon their resources immediately begins, and any little outlay at this juncture is abundantly compensated by its enabling these industrious emigrants the more quickly to push forward the furnishing of their new home. Clean combs from hives that may have lost their bees are readily accepted, and cause a great saving in time and material to the bees; these combs may easily be fixed by cutting them the proper size to fit within the frames, and making them firm by tieing with tape or fixing them with pliable wire. Artificial comb (see page 70) is often used, and has this advantage, that the combs are certain to be straight and regular, besides the saving in material to the bees.

These preparations must be made prior to the bees being hived, so that when a hive is so prepared, a swarm may begin to adapt whatever advantages they find ready for them; and it is truly marvellous what a swarm will do when thus furnished with combs in their new habitation. In these the queen can immediately begin to deposit her eggs, and the workers to store their honey, without having to wait for the construction of combs, which is a laborious occupation for the bees.

In some cases, fine white combs may be taken from the stock hive; the end frames are always the most free from brood. Care must be exercised not to rob this part of the hive too much; one comb may perhaps be removed in the course of the season without impoverishing the bees, but it is not wise to take more.

PUTTING ON SUPER HIVE.—A colony established a year or more is called a "stock," by way of distinction from a swarm of the present year. Supposing the hive to be a stock, the super should be given them at the early part of the season, say, if fine and warm, at the latter end of April or beginning of May; if the weather be then unfavourable, it is better to delay doing so until a more genial temperature. If the colony be a swarm of the present year, two weeks should be allowed to elapse from the time of tenanting a hive, before putting on the super; this delay is necessary to give the bees the opportunity of building combs in their new domicile, and of getting a store of honey for themselves before working for their master.

When it is wished to use a super, the crown board or roof of the stock hive must be taken away, the thin adapting or honey board taking its place; the two long slits at the sides are to give admission to the super. The bees will begin sooner and work faster if the eight bars are each furnished with artificial comb (as described page 71). We have had depriving hives very quickly filled when the bees were thus assisted. Combs that have been left unfilled may be fixed to the bars as before described; these must be white and clean, as dark comb should not be used for super hives. The combs when filled may be taken out singly, if desired for consumption, substituting an empty bar or comb; or should the bee-keeper desire to see a handsome super, he must wait until the bees have filled and sealed up all the combs, and then he may proceed to deprive them of it as before described.

TAKING OUT FRAMES WITH COMBS.—It is well for a beginner to practise the directions for opening and shutting up hives by using an empty hive, until he becomes familiar with the handling of the frames.

The first thing to do is to loosen the crown board or lid with a knife, drawing a piece of string underneath it to divide the wax or cement with which the bees make all secure. This string should be drawn through very slowly, so as not to irritate the bees. In hot weather, the crown board may be loosened by a lateral movement; but sometimes, for want of care, this loosening of the lid disturbs the bees, and as soon as it is removed, a number of them, enraged thereby, pour out and attack the operator. This and all other operations ought be done very carefully and gently; especial care should be taken not to prise the lid upwards, by way of wrenching it off, for the frames and combs are generally secured thereto, and there is a liability of rending the combs with it; this will greatly irritate the bees, and be otherwise injurious. When a hive of bees is enraged, there is little chance of pacifying them; it is best, under such circumstances to 'give in' at once, and not attempt to perform any operation, but to shut the hive up and beat a retreat, benefiting by the experience so as to do better in a day or so afterwards. There are various devices for intimidating or conciliating the bees, and one of these already spoken of is— smoke. So next time the experimenter makes his attempt let him raise the lid an inch or so, and blow a few puffs of smoke into the hive, which will cause the bees to retreat. This is best done by using our No. 12 fumigator, with a little of the prepared fungus lighted. Pipes or cigars are not convenient to use for this purpose when the head is enveloped in the dress. As soon as the lid is removed, a few bees will fly out to learn the cause of such an interference. Conciliation should then be introduced by having at hand a little sweetened water, which may be sprinkled, or rather let drop from a feather or brush. The sudden motion of the hand required in the act of sprinkling irritates the bees, so that instead of making them our friends, they may become our foes. Mr. Langstroth recommends that a fine watering pot filled with sweetened water, be used for the purpose. Care must be taken not to drench the bees; only just sufficient should be given to run down the sides of the combs as well as sprinkling the top. As soon as the bees really understand that syrup is being given them, they feast upon it instead of angrily attacking the operator. Thus pacified, and with gentle treatment, but little difficulty will be found in proceeding with the work required. But the unskilled operator should on no account neglect to put on a bee-dress and gloves as described above. We would err on the side of caution, although there is an old saying that "a cat in gloves catches no mice," and the apiarian will find that his fingers are not so free to work as he would like, for gloves make them rather clumsy in drawing up the frames.

The frames must now be gently prised from front to rear; this may be done with a small screw-driver or other stout instrument with a wedged end to go into the notches. The frames fit loosely so as to allow of a little movement from back to front; a lateral or side-way movement might kill the queen; or if not so fatal as that, might crush some of the bees and injure the brood combs, and must be carefully avoided. It, of course, depends upon what the operation is that has to be performed whether or no all the frames should be thus loosened. If it be for making artificial swarms, or for any purpose requiring an interview with her majesty, the whole of them must be loosened, because it often happens that all the combs have to be examined, sometimes twice over, before she can be discovered. Bees are very apt to build their combs in a slightly waving form, and in extracting one it will be needful to make room both for the comb and bees upon it to pass without scraping the next comb, and there will be a difficulty if the apiarian attempts to draw out one comb whilst the other frames are located in their appropriate notches. Let the operator gently proceed to lift say the third frame (allowing it to lodge on the little block that divides the notches) a little nearer to the fourth frame, and the second nearer the third, so as to admit of sufficient space to lift out the end one. Very carefully and slowly he should lift the frame by taking hold with thumb and finger of the projecting shoulders that rest in the notch; and he must not let it touch or scrape the next frame or the sides of the hive so as to crush or irritate any bees.

After the end comb is thus removed, it will be easy to extract the others, as there will now be plenty of room for drawing them out. A hive of exactly the same size should be at hand; and in case it be desired to remove the combs and bees into another hive, care should be taken that each comb occupies the same relative position that it did in the old hive.

In handling the frames, bear in mind that they are to be held perpendicularly. To gain a view of both sides of the comb when searching for the queen, or for any purpose requiring full inspection, with a little dexterity in twirling the frame round, the reverse side may be brought to face the operator, without letting the comb break away by its own weight and so fall out of the frame, which it will do if allowed to deviate from its upright or downright position. If the operator could see an experienced person perform the operation, he would quickly understand how combs may thus be handled without any risk of a smash.

When placing frames in the hive, care must be taken not to crush a bee between the projecting shoulders of the frame and the rabbets or notches on which they rest, and on no account must the frame be let down with a jerk, or the bees will become exceedingly fierce: the frame should be so slowly deposited in its place that a bee on feeling the slightest pressure may have the opportunity of escaping unhurt thereby. The crown board should be replaced by first resting its front edge in its place, and then slowly lowering the after part, looking carefully under, and momentarily raising it when necessary to avoid crushing a bee. Should the hive have its super on, the same directions may be followed. The super with its honey board may be bodily taken away, and so placed and confined for a time that robber bees cannot find an entrance, and also be far enough from the apiarian to be out of danger of being broken or overturned by him.

ADVANTAGES OF BAR AND FRAME HIVES.—It will be asked why all this trouble about bar and frames with straight combs built upon them? We have shown the full command which the bee-keeper has over a hive so constituted, and we now proceed to show how in skilful hands these advantages may be used successfully; though, in the hands of the unpractised and unskilful the contrary may be the result.

All the bars and frames in an apiary ought to be of precisely the same dimensions, so as to fit every hive. This is essential for the strengthening of weak hives. A hive that is weakly may often be advantageously strengthened by having put into it a comb of brood from a populous stock, to which an empty frame from the weak one may be given; no bees must be on the brood-comb—these should be shaken off or gently dislodged with a feather into the hive from which the comb is taken. The frames of the hive from which the comb has been taken, should be, one by one, placed so as to fill in the vacancy, leaving the empty frame nearest the side. When a hive has been in use many years, the combs become very black, and every bee that is bred in a cell leaves a film behind. It may be understood how in this way the cells become contracted, so that the bees that are bred in them are correspondingly reduced in size. After the lapse of say five years, it may be necessary to begin to remove the old combs. This may be done by cutting away the comb or by substituting an empty frame for one with old black comb, gradually moving the frames towards each other. By taking two away in this manner in the spring or summer of every season, the combs in course of five years may all be reconstructed and fresh clean ones be secured for breeding in, instead of the old black ones that otherwise would remain as long as the stock could live in the hive.

ARTIFICIAL SWARMING.—Every bee-keeper knows the anxiety there is in watching and expecting a swarm to come forth, fearful lest his favorites should, "like riches, take wing and fly away," a mis-chance that it is desirable to prevent. In our description of natural swarming this will be found fully treated of; we propose here merely to point out how, with the moveable frames, this work of nature may be assisted—we say assisted, because artificial swarming should, as nearly as possible, resemble natural swarming, that is, it should be performed at the same time of the year, and when the populous state of the hive makes a division desirable. This is easily known to be the case when bees hang out in clusters at the entrance, wasting their time in enforced idleness instead of being abroad gathering honey. It is also necessary that the hive contain drones.

When such is the state of the hive, the facility of forming an artificial swarm with a moveable frame hive is a decided advantage. The best time for performing the operation is about ten o'clock in the morning of a fine summer's day. The following directions should be carried out:—place ready a counter or bench that is firm and strong, and which has space on it for the inhabited—or rather the over-inhabited—frame hive and the empty one, which is about to be made the receptacle of a separate stock. The operator having on bee-dress and the other appliances ready, may now open the hive as before described, and proceed to take out the frames, carefully examining both sides of each comb to find the queen: she is generally in the centre of the hive, so that it is not always needful to take out all the ten frames. As they are examined, the frames may be put into the empty hive, and when the object of the bee-master's search is found, he must carefully remove the frame containing her majesty, and may place it temporarily in the empty hive at one end by itself. Next he must proceed to put the frames back into the old hive, closing up the vacancy caused by the removal of the comb with the queen on it, and leave the empty frame at the end. Then he may place the frame containing the queen—with the few bees that may be upon it—in the centre of the empty hive; then putting all the other frames in, and replacing the lid, the bee-master will place this hive in the exact position occupied by the old stock. The bees that are on the wing will go to the old spot, and finding the queen there, they will rally round her and very soon form a sufficient number to constitute a swarm; comb building will at once begin, the frames will in a week or so be filled, and a satisfactory stock will thus be established. By doing this at the right time, just before the bees are about to swarm, or when there are many drones, all the trouble of watching and waiting for them is saved. Mr. Woodbury claims the honour of having originated this mode of artificial swarming; subsequently Mr. Langstroth and others have described a similar process, no doubt originally, as far as they knew, therefore still more useful to us as a confirmation of the value of Mr. Woodbury's discovery.

This operation we performed, exactly as described above, with one of our improved cottage hives one afternoon at the latter end of May, 1862. Whilst inspecting our bees, we caught sight of the queen on the comb in one of the bell glasses. This was a chance not to be missed, and we immediately resolved to form an artificial swarm, for the hive was very full of bees. Besides, being obliged to be away from the apiary most of the week, we were glad of the opportunity of so easily establishing a colony without the uncertainty and trouble of hiving a natural swarm. In the first place we slid a tin under the bell glass, and removing the stock hive from underneath we took it a few feet away; then we placed an empty improved cottage hive where the old stock had stood, and put the glass of comb containing the queen and a few bees over one of the holes in the crown of this new empty hive. The bees that were left abroad, belonging to the old stock returned as usual to their old entrance as they supposed; soon a sufficient number formed a large cluster in the hive and began comb building, the queen remaining in the glass until the cells below were sufficiently numerous for her to deposit her eggs in them. The division answered exceedingly well, both hives prospered; the old hive either had some princesses coming forward to supply the loss of the queen, or the bees used a power that they possess of raising a queen from worker brood in the manner we have previously described.*

The foregoing account illustrates the successful formation of an artificial swarm; but with a cottage hive it was quite a matter of accident to have been able to get possession of the queen. With a a moveable frame hive she can at any suitable time be found.

Precisely the same plan is to be adopted with the old stock in the frame hive as we have described in the case of the cottage hive, that is to remove it some few paces off: when the hives are in a bee-house, a similar result may be attained by placing the new swarm for a day or two to the entrance used by the bees when with the old stock, and the old stock may be removed to an approximate entrance. Some apiarians recommend that a space be left between the two hives, by placing the hives on the right and

* See section 1, page 5.

G

left of the old entrance, in order that too large a proportion of
bees should not enter the new hive at the old position to the
impoverishment of the other.   But we have found the mode
adopted with the cottage hive answer so well, that we see no reason
for recommending any different plan.

It is the office of the bee-master to assist, not to go in the least
degree contrary to nature.   We know that when a natural swarm
issues forth it has its impregnated queen, and when located in a
new abode commences building worker combs, and leaves the
building of the few drone combs to a later period ; but if a division
of the hive should be made by putting *half the combs* in one hive
and half in another, the hive containing the queenless or embryo
queen will busy itself with building only drone comb, thus a number
of receptacles for useless bees is provided, which tends to weakness,
and eventually to loss of the hive.

In the plan we have recommended for forming two separate
families, we nearly follow the natural state of things ; the comb that
the queen is upon is the only one that is taken from the hive, and
this vacancy should be filled in by moving the frames together
so as to leave the empty frame at the end.   The bees, under the
government of the impregnated queen, construct the combs and furnish
their new abode, which, as before stated, they will do with worker cells.

By adopting the plan above described, the moveable bar and
frame hive will prove far superior to any of the dividing hives,
which provide for equal division of the combs.

Perhaps the greatest advantage the moveable frame hive
possesses is, that a full knowledge can be attained of its exact
state as regards the queen, the population and the quantity of food
in stock.   During weather of a genial temperature, the combs may
on any fine day be inspected, and thus a knowledge being gained
of the deficiency existing in a hive, the necessary means may be
adopted for supplying the want.   Sometimes such an examination
will verify the fears of the bee-keeper, when, having observed that
his bees have ceased to carry in pollen, he has thereby received
warning that the queen has been lost at some juncture when no
successor to the throne could be provided.   Such a hive has entered
on a downward course, and will dwindle away entirely unless a
queen should be given to it, or else, some combs containing
young brood not more than three days old.   By the latter method

the bee-keeper will gain an opportunity of seeing the bees set about their wonderful process of raising a queen from the brood thus provided for them.

When a bee-keeper has become skilful in his calling, he may be desirous to encourage the breeding of queens, or rather of preventing their destruction. He will seek to use the propagating instincts of the worker bees as a set off against that innate hatred of rivalry which prompts the reigning queen to kill the tender royal brood.

Hives found to be queenless may be supplied either with matured queens or with queen cells. If the latter are sufficiently numerous, their introduction may easily be effected by exchanging a comb in each hive; if they have to be cut out and placed loosely in the new hive, a triangular piece of comb should then be removed with them, to be used as a block in preventing any pressure coming on them. A space must be cut out of the middle in the centre combs of the hive into which they are to be introduced. Special care must be taken not to bruise the royal embryos, as they are particularly sensitive to pressure.

A very great advantage that the Woodbury bar and frame-hive possesses, is the safety and convenience with which a stock of bees can in it be transported to any part of the kingdom; and, by a few additional arrangements, stocks have been sent in it to distant countries. In many districts hives are removed to moors and heaths in autumn, for the purpose of gathering heather honey. In this operation the frames are a great support to the combs, very much lessening the risk of a break down and consequent loss.

From a hive that has been inhabited all the winter, we have not unfrequently lifted out the frames and removed the stock to a clean hive, and we believe that the change has always been useful. The bees find a clean floorboard and a clean hive to breed in, free from insects that may have harboured in crevices about the hive. When the change has been made, the old hive can be thoroughly cleaned, and used in the same way for making the exchange with another stock. The process for handling will, of course, be the same as before described. We have found that where this plan has been carried out, that the bees seem to progress faster. Perhaps a little stirring up may be useful in arousing them from the winter doze; the time we recommend for doing this is the beginning of April, but a fine warm day should be chosen.

## DRIVING.

Driving is an operation by which bees are induced to vacate an old settled hive and to enter an empty one. Many apiarians prefer this mode of effecting an exchange of hives to the plan of fumigating the bees.

The greatest success attending such a transfer will be in the case of hives well filled with combs that are worked nearly to the floorboard; and it may be remarked that bees are generally so far provident, that they leave an open space in which to pass underneath their combs over all the floor of the hive. When the old hive is inverted, the bees crawl up the combs, and thus more easily pass up into the new hive, which the operator places over the old one with the intent that they should enter it.

The best time for performing this operation is about the middle of the day, and when the weather is warm. It is essential that the operator be protected with a bee-dress and gloves, as before described; and previous to commencing his task, he must provide all necessary implements. These are:—a couple of hives, one of which should correspond in shape and size with the hive from which the bees are to be driven; a cloth to tie round at the junction when the new hive is placed on the old one; some string to keep the cloth in its place; an empty pail to receive the top of the old hive, if one of the old conical shape, but if the stock of bees is in a square box hive with a flat top, a firm stool will be the best; and a No. 12 fumigator with some fungus, which will complete the material of war. The bucket or stool must be placed securely on the ground about a yard from the place where the full hive stands; then a few puffs of smoke being blown in amongst the bees, will cause them to retreat up amongst the combs. The bee-master will now turn the hive* upside down very gently, letting it rest in the pail or on the stool; he then quickly places the empty hive over the full one, and ties the cloth round it to prevent any escape of the bees. If the cloth be damped, it will cling the closer to the hives. The third hive is intended to be placed on the stand formerly occupied by the stock, so as to retain the few returning bees which had been absent in the

* Care should be exercised in turning the hives over to keep the combs vertical, or they are likely to break from their foundations.

fields. Care must be taken that all crevices through which it is possible for the bees to escape from the united hives should be effectually closed. When the two are fairly united, the operator will proceed by rapping the full hive gently with the hands or a couple of sticks, more particularly on that side where the combs are the most thickly placed---that is, if the hive be not equally filled.

It generally happens that in about fifteen minutes the bees regularly commence the ascent; their exodus will be known by the distinct rushing sound which is always noticed when a colony of bees is on the move. The first thing bees do when disturbed is to fill their honey bags, as they invariably do at swarming time, consequently after the first rush into the new hive is over, as in the case of a swarm, the "flitting" bees are not much disposed to take wing. When the noise made by the ascending bees has been heard, and has in good degree subsided, the cloth may be removed, and the old hive, now deserted, may be taken indoors, and if a few bees yet remain, they may be brushed off with a feather. An experienced apiarian on first hearing the rushing noise before mentioned, will not hesitate to tilt the top hive over a little on one side, so that he may watch the bees during the ascent; the queen may be seen passing up, and if the operator desires to take her away, he can secure her by placing a wine glass over her. This expedient is often resorted to in the autumn when stocks are to be united, for in such a case the removal of the queen prevents some fighting.

If the taking of the honey be the object of the bee-master, then "driving" is manifestly a better plan than to resort to the fumes of sulphur for the purpose; for the bees from whom their store is taken, can be joined to stocks that are weak in numbers, with considerable advantage to the future prosperity of the apiary.

When the removed bees are to be joined to another stock, the operator will proceed as follows :—At dusk, dislodge the bees on to a cloth, sprinkle them with sweet syrup, and place the hive to which it is intended to join them, over the mass; they will gradually ascend into the hive placed for them, and early next morning the hive, with its slender stock thus augmented, may be removed to its stand. Should the operator not have been successful, or not sufficiently skilful to gain possession of the queen, he may leave it to the bees themselves to decide which queen they will have.

By this plan of "driving," artificial swarms may be secured by an "expert" even in common hives, though those do not afford the facilities for such a purpose as do the bar, or bar-and-frame, hives.

## CHANGING OLD STOCKS TO NEW HIVES.

We frequently find that the possessor of a stock of bees in a cottager's *common* straw hive is desirous of removing the whole stock of bees and comb into one of our improved hives, in which the honey may be obtained without the destruction of the bees. We mostly discourage such a transfer, attended as it is with much labour, and requiring a considerable amount of apiarian skill. An old fashioned hive may very readily be rendered a humane one, simply by cutting out with a sharp pointed knife the middle of the top of the hive; a piece may thus easily be taken out, so as to leave a round hole two or three inches in diameter, but care must be taken that the knife does not penetrate much below the straw, lest it reach the comb or the bees—and it will be safer for the operator to have a bee-dress on. There should be ready a round adapting board, with a corresponding hole, which may be secured on the top by putting four long nails through the same number of holes in the board; then a cap hive or a glass may be placed on the top, for the purpose of admitting the bees, who will soon crowd therein to work.

This hive or glass will form a super or depriving hive, and can be worked as profitably as most of the improved hives. For the sake of an improved appearance, an outside case, either of zinc, straw, or wood, may be dropped over all, and then, if well painted, the whole will form no disfigurement to any flower garden.

This is beyond doubt the easiest way of overcoming the difficulty, but as it may not satisfy all, we now proceed to describe how a complete transfer may be made. No hive offers such facilities for the correct placing of the combs in a perfectly upright position as does the bar and frame hive. As before remarked, we should be slow to recommend any one to attempt the operation who is not already pretty well accustomed to the handling of bees and acquainted with their habits; but by carefully carrying out the following directions any one may successfully perform the feat. The first thing is to get the bees away from the combs; there are two ways of doing this, one is by fumigation (see page 66), the other by driving (see page 84). Whichever plan may be resorted to, have the bees confined in the old hive on their stand until you are quite ready to admit them into the bar and frame hive. Have in readiness all the necessary appliances : these consist of a large knife for cutting the hive, a good sized table on which to lay the brood combs, a basin of water—for washing off honey which may besmear the hands,—tape or cotton string to fasten the combs in their frames, a pair of honey cutters for cutting out the combs, jars to hold the honey that runs out, and a feather for brushing off any bees that may remain. It is necessary that the operator should have on his bee-dress, and India-rubber gloves. Then begin by slicing off a piece of the hive with a sharp table knife; carefully cut out the working combs—cut them large, so that they will squeeze into the frames; and to be more secure, wind some of the tape round to keep them in position until made fast by the bees. After a few days, these fastenings may be removed. Care should be taken that the combs occupy the same position in the frames as in the hive from which they were extracted. Having thus prepared the hive, the bees may be let into it. It may be as well to keep them confined a few hours, giving them water at the top, until they make the combs secure; the new hive will then be less likely to offer an attraction to bees from other hives who, if feloniously inclined, might come to rob. This transfer should be made when the weather is such that the bees can fly about ; when not warm enough, it should be done in a room at a temperature of about 70 degrees. An expert apiarian could perform the operation in less than three quarters of an hour, and with little loss. A week or so after a swarm has left the old stock is perhaps the very best time for such a removal. Should the operation be performed in the open air, the bees from surrounding hives will be sure to come in great numbers to obtain a share of the honey necessarily exposed, for they delight in plunder. In order, therefore, to avoid annoyance to the operator, and the excitement which is certain to be induced in surrounding hives, it is better to conduct the dissection in some building with closed doors. In some instances a routing of this kind has a beneficial effect ; old stocks of hives that have previously appeared to be dwindling, are often aroused to activity by their removal into a fresh domicile.

## WEIGHING HIVES, &c.

One of the most effectual modes of ascertaining the condition of a hive is by weighing it. Such knowledge is most important at the close of the gathering season, in order that the bee-keeper may determine whether he ought to give his bees artificial food to enable them to live through the dreary winter. A knowledge of the numerical strength of the colony is also useful, in enabling the bee-keeper to decide which hives will be benefited by being joined together, which may be done on the plan explained in the articles on " Fumigation " and " Driving."

A hive can very easily be weighed, if a *Salter's Spring Balance* be suspended near the apiary. The hive, having a strap or cord passed under and over it, and crossing at right angles on the top, may be hooked on to the balance, so that the weight will be indicated on the dial. The weight of the hive should be marked on it when empty, so that the exact amount of its contents may at any time be ascertained. A colony of bees at Michaelmas ought to weigh from 20 to 24 pounds, that is, exclusive of the hive ; if falling short of that weight, the hive should be made up to it by the artificial means before recommended.

Experienced apiarians are able to judge of the weight of a hive by lifting it a few inches from the stand ; or by looking in at the windows of a stock hive, a conclusive opinion may be formed as to the state of the colony. If the combs within view be well filled and sealed, it will be safe to consider that the hive contains sufficient stores to carry the bees through the winter.

## V. MISCELLANEOUS INFORMATION.

### STINGS : THEIR PREVENTION AND CURE.

Some of our readers may deem us neglectful in having, as it were, left them to struggle through their bee-keeping novitiate without informing them how to avoid being stung by their docile but well armed flock. Of course, having described the bee-dress, we have supposed that the apiarian was clad, if not " in complete steel," at least, in the head-gear and gloves, which will render him invulnerable.

The best safeguard from the anger of bees—as indeed from the malice of men—is a quiet and peaceable spirit.  The apiarian will learn to handle his bees not only as " if he loved them,"—as the quaint angler says—but as if he fully believes that the bees love *him*. This they will do whenever he approaches and treats them gently. There are some cases of exception to this generally peaceable disposition of the bee ; sometimes a few bees are dyspeptic, and refuse to be pacified—let their master seek to bribe them never so wisely.   Then, too, sometimes the bee-master himself may be dyspeptic, which the unerring olfactory sense of the bees speedily detects, and their anger is immediately aroused.  Some few persons, owing to constitutional peculiarities in their breath or insensible perspiration, are objects of constant animosity with bees, who by driving them from the apiary, are giving a physician's advice without charge for a fee.  Some of the choicest perfumes used by ladies are offensive to bees : and one may feel very certain that the " fine puss gentleman," who disgusted the brave Hotspur with his " pouncet box " and praise of " 'parmaceti for an inward bruise," would have been speedily driven from an apiary in ignominious flight. Occasionally, even a skilful apiarian may inadvertently crush a single bee ; such a mischance is detected by the community with much more facility than by any " crowner's quest," and their prompt verdict decrees the summary punishment of the offender.   There would be much less fear of stings if it were always remembered that bees are never aggressive.   " Defence, not defiance," is their motto.  They scarcely ever attempt to sting when away from the hive, and very seldom indeed at the time of swarming, for then they are gorged with honey.   When molested by angry bees, the safest and best retreat is a green bush.   Thrust your head into this, and the bees will soon leave you.

Yet some people appear to think they must inevitably be stung if they meddle with bees ; and for their sakes it is needful to explain why it is that a sting is painful, and how the wound inflicted by the bee may be cured.   Those familiar with the usual microscopic objects will know how marvellously delicate, and yet effective, is the mechanical structure of a bee's sting.  This weapon as we see it with our naked eye—finer than a needle's point—is only the sheath, which lengthens or contracts like the tubes of a telescope.  From the sheath is projected the dart, which is double,

each half of it piercing alternately deeper into the wound made by
the sheath.  The dart is barbed on each side, so that the bee when
*very* angry is scarcely ever able to withdraw it :—

> " Deems life itself to vengeance well resigned ;
> Dies on the wound, and leaves the sting behind."

If the patient who receives the sting could only take it patiently,
it would not prove half the inconvenience to him that often is the
case.   There are indeed some happy mortals whose " blood such an
even tenour keeps," that a bee-sting is to them simply a puncture,
and nothing more.   Dr. Bevan has suggested that lovers should
subject themselves to the ordeal of a bee-sting, in order to prove,
we suppose, that their temper is proof against "the *stings* and
arrows of any outrageous fortune" that matrimony can bring.

It is the homœpathically minute tincture of poison injected by the
bee which causes inflammation.  The first thing to do is to remove the
sting, which, even when detached from the bee, will continue to penetrate still further into the wound.  Next, press the hollow point of a
watch-key exactly over the place stung ; this will express a considerable portion of the virus.   Then dip the hand or bathe the part with
cold or tepid water, for the poison is volatile, and will thereby be
dissipated to a great extent.  On no account whatever should the part
affected be rubbed ; to do that will diffuse the poison, and increase the
inflammation.  The specific remedy for a bee-sting is taught us by
chemistry : the venom is an acid, which an alkali will immediately
neutralize when brought into contact with it.  Spirits of harts-
horn will generally be found effectual for the purpose, and should
always be kept in an apiary.  There are also several other remedies,
more or less effectual, according to the special constitution of the
patient.  A strong infusion of tobacco water applied to the wound
after the sting has been extracted, is a specific for many persons ;
others find relief from the application of a sliced onion.

We have heard the remark from several who have kept bees for
years, that the poison from a sting has little or no effect on them ;
after receiving many inflictions, their flesh appears to become
so little affected that the swelling and pain at one time experienced
no longer trouble them.

## POLLEN; OR, FOOD FOR INFANT BEES.

Bees, when fully grown, feed almost wholly on honey; but the larvæ require for their development a more substantial kind of nourishment. Such solid fare is found by the bees in the *pollen* of flowers, a farina which contains some of those nitrogenous elements in which honey is deficient. The body of a worker-bee is covered with hairs, to which the pollen adheres when, by contact with the bee, it is rubbed from the anthers and stamens of flowers. The bee with its fore legs then brushes it off, and moulds it into the pellet shape suitable for carrying it in the "baskets" or grooves on its thighs. Dewy mornings or humid bowers suit the bees for the gathering of the pollen. If the atmosphere be too dry for kneading it into pellets, they roll themselves in the blossoms, and trust to the good offices of the bees at home, who, on their return, brush off the farina into the cells intended for it. A portion of this "bee bread" is taken at once by the "nursing bees," which are supposed to subject it to some change before offering it to the larvæ; but the greater part of the pollen is stored away, and sealed over in the cells for future use. In April and May the bees are frequently busy "all the day" in gathering pollen, and often one community of bees will collect about twenty pounds weight of "bee bread" in one season.

One of the objects of the apiarian is to assist the bees in providing for the nurselings of the hive. A German pastor, HERR DZIERZON, first suggested the plan of providing the bees with "unbolted rye meal" as a substitute for the farina of flowers. He had observed that, in early spring before the flowers were open, his bees had entered a neighbouring corn mill, from whence they returned laden with rye flour. Since his discovery, most bee-keepers in early spring place either rye or wheat meal near the apiaries; to this artificial store the bees repair by thousands, and seem to rollick in the enjoyment of so much plenty, many of them returning to the hive as dusty as millers. The object in thus supplying them is, that the brood may be rapidly brought forward and early swarming induced. In this way, a few pounds of rye meal at one penny per pound may tend to the production of very many pounds of honey of twelve times the price.

In gathering pollen from flowers, bees are doing more than merely providing for their own community. Whilst humming through our gardens they are assisting to propogate our flowers, and their merry buzz in our orchards indicates that the blossoms of spring will in autumn fulfil their promise by abundance of fruit. In Mr. Darwin's remarkable work, "The Fertilization of Orchids," the mystery of the fructification of flowers is scientifically explained ; but before the subject was so fully understood, it was quite believed that bees in passing from flower to flower performed some important service. Owners of fruit trees have noticed, in a season generally unfavourable for the orchard, that if during only one fine forenoon the bees had spread freely amongst the blossoms of a particular tree, that it would prove more fruitful than its fellows. On this account the orchard is a good place for an apiary, for it seems— more abundant the honey, more plentiful will be the fruit. Bees bear the fructifying matter from one *sex* of flowers to the other, but they confine their attention to one *kind* of flower during each excursion : and the careful observer may see how the colour of the pollen on the bodies of the bees will vary from yellow to red and brown, according to the kind of flowers from which it has been gathered. The gathering of pollen, its use by the nursing bees, and the storing of it in the cells, afford to the bee-keeper opportunity for observations of exceeding interest.

## PROPOLIS ; OR, BEES' CEMENT.

The old notion that wax is gathered by bees from flowers as they gather honey, has long since been set aside by the discoveries of Hornbostel and Huber. Wax is an oily substance secreted from the honey in the bodies of the bees, on which it forms in thin flakes. But there is "a resinous substance, very tenacious and semi-transparent," which is indispensable for the bees as a cement wherewith to fix their combs and fortify their hives against intruders, and this is "propolis." The bees, in working the propolis, often soften it by blending it with a portion of wax; but they have to extract it in its natural state directly from the bark and buds of certain trees. The bark of the willow, the leaf buds of the poplar and alder, and the unopened blossoms of the hollyhock, are very usual sources of propolis. In the case of a new swarm, as bees must have this glue before they can begin to build their combs, they will resort to most unlikely places to obtain it. Sometimes they will enter a paint shop and attack the varnish, and it is said they have been seen to obtain propolis from the pitch and rigging of a ship. These circumstances afford intelligible hints to the apiarian, who, if his bees have not easy access to firs, poplars, or willows, will provide some glutinous or resinous matter which may serve for a substitute. The extracting of the propolis costs the bees very considerable labour, which they should be relieved of as much as possible, in order to facilitate their great work of honey gathering. Bees choose the warmer part of the day during which to gather propolis, as then it does not so rapidly stiffen. Frequently when they arrive at the hive it has become so hard that the other bees are scarcely able to gnaw it from their thighs.

With propolis bees fasten down their hives, stop up crevices to exclude moths and ants, and sometimes use it to narrow the entrance of their hive against the invasion of wasps. Extraordinary anecdotes are told of the prompt and ingenious use they make of this substance. Reaumur relates that a snail having been observed by the bees on the window of the hive, they proceeded to glue the shell to the glass, and there sealed down the intruder in hopeless durance. In another case, that of a slug, or snail without a shell, the bees having slain it with their stings, were quite unable to remove it from the hive. With wonderful foresight, they then proceeded to secure their community from the noxious effects likely to arise from the decay of the carcase; and this they did by completely enveloping it with a coating of impervious varnish. Huish relates a similar occurrence in the case of a mouse caught in a hive by bees. Propolis yields *benzoic* acid, and contains some aromatic properties.

## PASTURAGE FOR BEES.

" Bees work for man; and yet they never bruise
Their master's flower, but leave it, having done,
As fair as ever, and as fit for use "

Apiarians generally agree in the opinion that very little can be done in the way of providing any special forage for bees. Yet bee-fanciers are always interested in observing which the flowers are

that the bees prefer; and there are certain well established conclusions
as to the kind of district and season which are the likeliest to
produce a good honey harvest. There is an old saying that a
country which produces the finest wool also yields the best honey;
and a pastoral district is decidedly better than one under tillage.
The principle of the matter is, that the bees are best suited with a
long dry season—an early spring, a hot summer, and a late autumn.
As not one of these blessings can be commanded by the apiarian,
his art must be applied to provide some mitigation of the injury
suffered by the bees when the season is short or wet. For early
spring, the crocus, the blue hepatica, and the violet, all afford good
supplies of honey, and if cultivated near the apiary, will be of great
service when the wild flowers are backward. All varieties of the
willow and poplar furnish early supplies of honey, as well as of the
propolis of which we have spoken; the blossoms of the gooseberry
and currant are very useful for the bees in May. Wet, when it
enters flowers of any kind, prevents the proboscis of the bee from
reaching the secret source of honey. On this account, it is well
to know, as does the bee, that the drooping blossoms of the raspberry escape the effect of the showers, and honey is gathered from
them when other flowers are drenched within as well as without.
For a similar reason, the borage (*borago officinalis*) is valuable for
bees; and, also, because that plant continues to flower until the frosts
set in. The honey both from raspberry blossoms and borage is very
superior. Mr. Langstroth says that " the precipitous and rocky
lands of New England, which abound with the wild red raspberry,
might be made almost as valuable as some of the vine-clad terraces
of the mountain districts of Europe." The " golden rod," and also
asters, afford superior honey for autumn gathering. Dzierzon
strongly recommends buck-wheat being sown in the winter stubbles
on behalf of the bees, and he tries hard to persuade farmers that it
is to their interest to cultivate it. It should be named that all the
ordinary fruit blossoms, especially those of the apple, supply
abundant store for bees.

It is, however, to wild or field flowers that the bee-master must
chiefly look for the raw material on which his myriad artisans shall
exert their skill. The white clover of the pasture—the wild thyme on
the hill—the heather on the moors—the furze and the broom on the
sandy waste—offer exhaustless stores for a greater number of bees

than can ever be located near them. There are also two or three peculiar sources of honey which one would not have suspected, as, for instance the blossoms of the onion plant, of turnips, and in still greater degree, the flower of the mustard plant. In those districts of England where mustard seed is cultivated so extensively, it would be well worth while for the farmers to keep large colonies of bees. Another, but a very uncertain source of honey, is the "honey-dew," which in some seasons appears in large quantities on the leaves of the oak, the lime, and some other trees.

It is important to mention that bees in the principal breeding season require a plentiful supply of water. Owing either to their carelessness or eagerness, they are frequently drowned when drinking from any large quantity of water; the bee-keeper should, therefore, place near the hives shallow vessels of water containing pebbles, on which the bees may alight to take frequent but temperate draughts.

## THE LIGURIAN OR ITALIAN ALP-BEE.

A new, or rather a re-discovered, variety of bee has recently been brought into practical use amongst apiarians in Germany and America, as well as in this country. The ordinary bee is the *Apis Mellifica* of naturalists; the new kind is the *Apis Ligustica*. It was also named "The Ligurian Bee" by the Marquis de Spinola, who found it in Piedmont in 1805; and he considered it to be the principal species known to the Greeks, who speak of the "best kind" of bee as being of a red colour. Leading apiarians agree in pronouncing these bees to be justly entitled to the high character given to them. Their special advantages are—greater fecundity of the queens, less irascibility, and a more handsome appearance, for being of a golden colour, they are prettier than our black bees.

Our own experience with the Italian Alp-bee enables us to corroborate the statements which have been made in favour of this new variety. We find the queens more prolific than those of the common kind, and the quantity of honey produced is greater. These two facts stand as cause and effect; the bees being multiplied more quickly, the store of honey is accumulated more rapidly, and the Italian bees consume, if anything, less food than the common kind. When of pure Italian blood, these bees are by some apiarians thought to be hardier than our own. That they forage for stores with greater eagerness, and have little hesitation in paying visits to other hives, we can testify from our own observation. The following anecdote will illustrate their intrusive propensities. Another bee-keeper living in the neighbourhood of our apiary, when inspecting our hives, observed the yellow bees; he exclaimed, "Now I have found out where those strange-looking bees come from; for," said he, "these yellow-jackets are incessant visitors to my hives. I thought they were a species of wasps that had come to rob, and until now I have been unable to account for their appearance at the entrance of my hive, so that I have killed them by hundreds." This was not at all pleasing intelligence for us, and we trust that our neighbour has been more lenient to "the yellow-jackets" since his visit. We are inclined to believe that more visiting takes place amongst bees of different hives than bee-keepers have been accustomed to suppose; where the Italian and black bees are kept near each other, the foreigners being conspicuous by their lighter colour, there is less difficulty in identifying them when at the entrances of other hives.

This season (1864) we have had more honey from a Ligurian stock than from any one of our colonies of black bees. From this Ligurian hive we have taken a glass super containing 40 lbs. nett of honey, besides having drawn from it an artificial swarm; and, after all, it remains the strongest hive in our apiary.

The Baron Von Berlepsch and Pastor Dzierzon, who are probably the two most intelligent and skilful bee-keepers of Germany, award to the Italian a very decided preference over the common bees. The Baron says that he has found:—1. That the Italian bees are less sensitive to cold than the common kind. 2. That their queens are more prolific. 3. That the colonies swarm earlier and more frequently. 4. That they are less apt to sting. 5. They are more industrious. 6. That they are more disposed to rob than common bees, and more courageous and active in self defence. They strive, whenever opportunity offers, to force their way into colonies of common bees; but when strange bees attack their hives, they fight with great fierceness, and with incredible adroitness."

It is said, that the Italian bee can extract honey from some flowers which the common bee is unable to penetrate. For instance, the blossom tubes of the red clover being too deep for the proboscis of the common bee, that flower is useless to them, although so plentiful; but, says Mr. Langstroth, the American apiarian, the Italian bee visits the red clover assiduously, and draws large quantities of honey from it.*

The introduction of this new variety of bee into England was through our agency. M. Hermann, a bee cultivator at Tamins-by-Chur, Canton Grison, Switzerland, wrote to us on the 5th July, 1859, offering to supply us with Italian Alp queen bees. This letter, or an extract from it, appeared in the current number of the *Journal of Horticulture*, (then called the *Cottage Gardener*,) a periodical that regularly opens its columns to apiarian subjects. Prior to this the *Italian Alp*, or as it has been named the " Ligurian " bee, was UNKNOWN IN THIS COUNTRY, except to a few naturalists. The letter referred to attracted the attention of that intelligent apiarian T. W. Woodbury, Esq., now so well known as ·the " Devonshire Bee-keeper." On the 19th of July, that is a fortnight after Mr. Hermann's offer, we received a consignment of Italian Alp bees, being the first imported into England. With these, Mr. Woodbury also received one queen-bee and a few workers, which he introduced into a hive of English bees from which the queen had been taken. His efforts were very successful, and " the spring of 1860 found him in possession of four Ligurianized stocks." His subsequent experience with the Italian Alp bee he has fully described in a communication to *The Bath and West of England Agricultural Journal*.

Subsequently Mr. Hermann sent us a copy of his pamphlet entitled "'The Italian Alp Bee; or, the *Gold Mine of Husbandry*," with the request that we should have it translated from the German, and that copies of it should be printed in the English language. The pamphlet was speedily published by us, and although singular as a literary production, it may be useful for the advanced apiarian.

Certainly the bees are partially of an orange or golden colour, and if one could believe the golden anticipations indulged in by Mr. Hermann respecting them, it would be sufficient to identify the Italian Alp bee as the species described by Hood in Miss Kilmansegg ;—those which dwelt in

* This opinion is not held by the closest observer of Italian bees in England.

H

" A golden hive, on a golden bank,
Where golden bees, by alchemical prank,
Gather gold instead of honey."

Poor Hermann himself did not profit by the discovery of the mine.
His enthusiasm for the culture of the Alpine bee instead of wafting
him to El Dorado appears to have driven him to the prosaic retreat
of a debtor's prison. By the latest accounts we learn that—his
ardent spirit chafing against his incarceration,—he had made an
attempt to escape, and in doing so, fell from the prison walls and
broke both legs. We understand that he has since died.

In the pamphlet referred to, Mr. Hermann gives the following
description of what he insists on designating as *Apis Helvetica* :—
" The yellow Italian Alp-bee is a mountain insect; it is found
between two mountain chains to the right and left of Lombardy
and the Rhetian Alps, and comprises the whole territory of Tessins,
Veltlin, and South-Graubunden. It thrives up to the height of
4,500 feet above the level of the sea, and appears to prefer the
northern clime to the warmer, for in the south of Italy it is not
found. The Alps are their native country, therefore they are called
*Yellow Alp-bee,* or tame house bees, in contra distinction to the
black European bees, whom we might call common forest bees,
and who, on the slightest touch, fly like lightning into your
face. (?)

" As all good and noble things in the world are more scarce
than common ones, so there are more common black bees than of
the noble yellow race, which latter inhabit only a very small piece
of country, while the black ones are at home everywhere in Europe,
and even in America."

Notwithstanding the death of M. Hermann, we shall in future
be able to supply all bee-keepers, who may wish to possess stocks
of the genuine yellow Alpine bees.

The Italian varies but little from the common bees in its
physical characteristics. The difference in appearance consists in
the first rings of the abdomen, except the posterior edge, and
the base of the third—being of an orange colour instead of a
deep brown. These orange coloured parts are transparent when
closely examined with the sun shining on them. The Italian bees
are more active than common bees when on the wing.

During the summers of 1859 and 1860 we had over from the

Continent a great number of Ligurian queens; these were sent to all parts of the kingdom. We regret to say that but few were successfully united to English stocks. It requires a considerable amount of apiarian skill to accomplish the union, so that we find by experience it is best to send out complete Ligurian stocks. This is particularly desirable now that the packing of whole hives is so easily accomplished by us with the aid of bars and frames. We have sent a great number of stocks to all parts by rail.

Mr. Woodbury, owing to his knowledge and skill in bee-keeping, was eminently successful in propagating the Ligurian bees first imported into this country; and we would recommend all who may be interested in the subject, to peruse the interesting articles written by him in the *Journal of Horticulture*. He has shown great patience and energy by his labours in the rearing of queens and the multiplication of stocks, for which he merits all praise. On the other hand, Mr. Woodbury has to thank the Ligurian bees for the development of his own skill, for although prior to this he was a valued correspondent to the before named Journal, his commencement of scientific bee-keeping must be dated from the time of their introduction.

We now add to the other testimonies already cited, that of Mr. Woodbury as to the superior qualities of the Ligurian bees. The following is extracted from a paper contributed by him to the *Bath and West of England Agricultural Journal* :—" From my strongest Ligurian stock I took eight artificial swarms in the spring, besides depriving it of numerous brood-combs. Finding in June that the bees were collecting honey so fast that the queen could not find an empty cell in which to lay an egg, I was reluctantly compelled to put on a super. When this had been filled with 38 lbs. of the finest honeycomb,* I removed it, and as the stock-hive (a very large one) could not contain the multitude of bees which issued from it, I formed them into another very large artificial swarm. The foregoing facts speak for themselves; but as information on this point has been very generally asked, I have no hesitation in saying that I believe the Ligurian honey-bee infinitely superior in every respect to the only species that we have hitherto been acquainted with."

* This super was exhibited at our stand in the International Exhibition of 1862.

After such decided testimony as this, corroborated, as it is, by many other observers, there seems every reason to expect that the Ligurian bee will gradually supersede the common kind throughout the United Kingdom.

## LIVING BEES AT THE INTERNATIONAL EXHIBITION OF 1862, SENDING BEES TO AUSTRALIA, &c.

The engraving represents our stand in the Agricultural Department of the International Exhibition of 1862. The space granted us in the World's great Fair was somewhat limited; but we were able to exhibit a tolerably complete stock of apiarian apparatus and all the more important bee-hives. Amongst these was a Unicomb Hive stocked with the Yellow Alpine or "Ligurian" bee. This was an object of great attention, and daily hundreds of visitors flocked round our stand in order to watch the movements of the Italian queen with her gay and busy subjects. The entrance way for the bees being in the "Open Court," to which all visitors had access, it was necessary to place the hive in an elevated position, so as for it to be beyond the reach of incautious passers by, and to obviate any chance of annoyance to the vast crowds of people continually around.

Among others who took a deep interest in our exhibition, was Mr. Edward Wilson, President of the Acclimatisation Society of Victoria. This gentleman requested us to pack four stocks of the Ligurian bees for conveyance to Melbourne. With the assistance of Mr. Woodbury—whose aid was, indeed, essential—these stocks were sent off on the 25th of September, 1862, by the steam ship *Alhambra*, so as to arrive at the colony during the Austral summer. The hives were Woodbury-frame hives, having ample space and ventilation, as well as the means of supplying water to their inmates during the voyage; there was, also, a sufficient store of honey to last until the following March. The bees arrived at Melbourne, where they were released after an imprisonment of seventy nine days, and have since rapidly multiplied, the climate and pasturage of Australia greatly favouring the increase of this superior variety of the bee.

Mr. Wilson was so well pleased with the careful manner in which these stocks were fitted out for their voyage across the seas, that he subsequently instructed us to prepare him three more hives, which were sent out in a sailing vessel. Owing to the mismanagement of the water supply during the voyage, only one stock survived in this instance.

Upwards of twenty years ago, we sent a Nutt's Hive stocked with bees to New Zealand. We then adopted the plan of fixing the hive in a meat safe, so that the bees could fly about a little, and also cleanse the hive of their dead, for bees are very attentive to sanitary arrangements; they always remove the dead ones from their midst, and do not void excrement within the hive.

When bees are shut up in their hives too long, they are apt to be attacked by a disease called by apiarians—dysentery. Sometimes when confined by the unfavourableness of the weather in winter or the lateness of spring, this disease produces serious mischief amongst the bees. Various remedies have been recommended; but we believe the best is to see that there is wholesome food within the hive, and plenty of it, and when fine weather returns, the health of the bees will return with it. Dampness of the hives, and too late feeding in the autumn, are also frequent causes of this disease.

## BEE-KEEPING IN LONDON.

There are many persons now in this noisy city pent, who frequently remember the days of childhood when, among pastures of clover or amidst flowery heath and woodlands, they listened to the cheerful hum of bees. Partly from a desire to revive those old associations, and also from a natural liking for the tendance of living creatures, such persons would be glad to keep bees if they thought it possible to do so in London or its suburbs with any chance of success. We do not wonder that many should doubt even the possibility of bees feeding themselves amidst such an "endless meal of brick;" but we can easily prove that bees, if not placed too near to smoky chimneys, are able to produce honey, both for themselves and for their masters. To make this plain, we will mention some special instances of metropolitan bee-keeping.

About ninety years ago, a Mr. WILDMAN kept a bee-house and honey warehouse near to Middle Row, Holborn. He was not only a tradesman, but was also the apiarian of his day. He kept hives of thriving bees on the roof of his house in Holborn, and many of the nobility and gentry used to mount thither in order to inspect the apiary. At that period, St. Pancras was a "village two miles north-west of London," and what is now the Regent's Park was open country. It was then much easier for London bees to find their favourite forage, but Mr. Wildman believed that his hives were filled with stores from a considerable distance. Whilst enjoying his country rambles on Hampstead Heath, he had a shrewd suspicion that many of the bees he there observed gathering honey were labourers from his own apiary. In order to identify his own flock amongst the rest, he hit upon a homely but very effective expedient. Having borrowed Mrs. Wildman's " dredging box," he stationed himself near the entrance of his hives, and gently dusted his bees with flour as they issued forth. He then betook himself to Hampstead, where he found his previous surmise confirmed, for there were numbers of his bees in their livery of white.

Wildman became noted for the remarkable control he obtained over his bees, many instances of which he exhibited before the public. Many of his operations with them were regarded as feats of legerdemain by the uninitiated, as when he appeared before the king George III. with a swarm of bees hanging in festoons from his chin, or suspended in a cluster at arm's length. The *Journal of Horticulture* recently, in alluding to Wildman, gives the following particulars as to his performances:—

Near the "Three Hats," Islington, was a place of popular entertainment called "Dobney's Tea Gardens," kept by Mrs. Ann Dobney. These gardens occupied the ground between White Lion Street and Winchester Place, and were established as far back as 1728. In 1771 the house was taken for a short time as a boarding school; but it was soon changed to its original purpose as a place of amusement, for in 1772 Daniel Wildman exhibited bees here. This is a copy of the advertisement:—

" June 20, 1772. Exhibition of bees on horseback ! at the Jubilee Gardens, Islington, (late Dobney's), this and every evening until further notice (wet evenings excepted.)

" The celebrated Daniel Wildman will exhibit several new and amazing experiments, never attempted by any man in this or any other kingdom before. The rider standing upright, one foot on the saddle and one on the neck, with a mask of bees on his head and face. He also rides standing upright on the saddle with the bridle in his mouth, and by firing a pistol, makes one part of the bees march over the table, and the other swarm in the air and return to their hive again, with other performances too tedious to insert. The doors open at six; to begin at a quarter before seven. Admittance:—Box and gallery, 2s.; the other seats, 1s.

The secret of Wildman's skilful manipulation is well understood now; it consisted in a careful holding and disposal of the queen, together with confidence in the generally inoffensive disposition of bees. Dr. Evans, whom we have often quoted for his correct information in apiarian matters, thus speaks of his feats:—

"Such was the spell which, round a Wildman's arm,
Twined in dark wreaths the fascinated swarm;
Bright o'er his breast the glittering legions led,
Or with a living garland bound his head.
His dextrous hand, with firm yet hurtless hold,
Could seize the chief, known by her scales of gold,
Prune, 'mid the wondering train, her filmy wing,
Or o'er her folds the silken fetter fling."

To recur to our subject. After the days of Wildman, our own establishment in Holborn became widely known for bee hives and honey. Although we never attempted to start a London apiary at all approaching in extent that of our predecessor, we have occasionally kept bees on the house top both in Holborn and Regent Street. At both those situations, we have noticed that the bees bring "pollen" as well as honey into their hives. Last summer there was brought under our notice an illustration of the acuteness of the scent of bees, and of their diligent search for food, proving too that if sweets can be obtained even from unusual sources, the bees will find them out. A poor woman who, at the corner of an adjacent street vends "brandy balls," "toffee," "rock," and other saccharine compounds—all well known to and appreciated by most juveniles,—used to receive frequent visits from our bees. Their visits to the old dame's domain were at first rather interesting, and if the few pioneers who had the sagacity to find such a store had kept the secret only to themselves, their company would not have been objected to. Such selfish policy does not, however, accord with the social instinct of bees, and these soon informed their companions of the good fortune provided for them in an archipelago of sugar islands. Day by day the swarms of these uninvited visitors increased, until all legitimate customers were beaten off; and the old dame had to see, not only her hope of gain destroyed, but her stock of "goodies" sensibly diminishing by the thefts of these brigands of the air. She could not, or dare not attempt to, drive the intruders away, so made diligent enquiry as to where the robbers were harboured. Having traced them to our establishment in Regent Street, she came to implore of us to move the bees if possible, or she would have to move her stall, and so lose her "connection" in the "toffee" and "rock" trade. Wishing not to hinder the poor woman in gaining her livelihood, we decided on removing our bees into the country.

It is difficult to assign an exact limit to the distance that bees will go in search of honey yielding blossoms. It has been proved by various experiments that they will fly say five or six miles, if the supplies are scanty within a shorter radius; but bees well understand that first of all economics,—the saving of time, and if they can find forage near at hand, they prefer it. Hence, other things being equal, the quantity of honey stored will be in proportion to the contiguity of good pasturage. In this way it is that the systematic removal of hives, as practised in many districts, has such a notable effect on the honey harvest.

During several years we kept bees in the Zoological Gardens, Regent's Park, and have there frequently taken full and handsome glasses of honey. The position of our apiary was on the site now occupied by the " New Monkey House." The Society promise to erect a new building for an apiary in the course of the ensuing summer. The visitors to the Gardens found considerable interest in watching the bees in our glass hives, and are now much disappointed at the absence of so entertaining an exhibition.

A gentleman residing in St. James's Place, has for some considerable time past kept bees in his garden there. He uses our improved cottage hives, and his bee-keeping is decidedly successful as he generally takes some fine glasses of honey each season, besides leaving sufficient as winter store for the bees. For a London situation, St. James's Place is a very favourable one; the gardens behind the houses pleasantly face the Green Park, so that the bees have an uninterrupted flight to start with. They are also within easy range of the richly flowered gardens of Buckingham Palace and those of the nobility and gentry who reside around the Parks. To those gardens the bees of St. James's Place resort, without waiting for any license or certificate from the royal and noble owners of the honey yielding preserves. Being within a short distance of our establishment, when this gentleman's bees swarm he generally sends to us for assistance in hiving them.

The neighbourhood of St. John's Wood, and indeed almost all the suburbs of London, are favourable for the production of honey. We mention St. John's Wood because, from the fact of having kept bees there ourselves, we are able to prove by our experience that they do well in that locality. We have several customers on nearly all sides of the town, who have each had this year a considerable surplus of honey in their "supers," after leaving sufficient for the bees themselves in the lower or stock hives.

We exhibited in our window last autumn a super of fine honey from the apiary of Shirley Hibberd, Esq., the proprietor and editor of the *Gardiner's Weekly Magazine.* It is a box containing 20 lbs. nett weight of honey, and was produced at Stoke Newington, only 3¼ miles from the General Post Office.

The *Times* "Bee-Master," whose letters from Tunbridge Wells have awakened so much interest in this pleasing pursuit, also commissioned us to exhibit a "super," produced under his own management in that locality. A friend of ours at Exeter had upwards of 400 lbs. of honey, of excellent quality, though one of his apiaries is quite within the city.

The last has been an excellent honey yielding season; our own bees, at Dorking, in Surrey, have produced us large quantities, and the accounts from nearly all parts of the country coincide in stating that the bees have in the year, 1864, enjoyed unusual opportunities for accumulation. In not a few localities, the season of 1863 was even more abundant.

## WASPS AND MOTHS.

Bees have few enemies more formidable than wasps. The most effectual method of checking their invasion of hives, is to have as narrow an entrance as the bees can do with. If a stock be not very weak in numbers, the bees will be well able to guard a small aperture, and can repel the attacks of those insidious and merciless robbers. On this account, the entrance to our No. 5 hive as described at page 31, may be used.

The bee-keeper is interested in preventing the increase of wasps; it is, therefore, a good practice for him to set a price on queen wasps in the spring, the death of one of them at that time being equivalent to the destruction of a whole nest.

Should nests be found in the neighbourhood of an apiary, their annihilation must be accomplished either by blowing them up with gunpowder, an operation well understood by most country lads; or any other effectual method. The late Mr. Payne recommended that a small quantity of gas tar should be put into the mouth of a wasp's nest, and if then covered with earth, the total destruction of the wasps will be accomplished without further trouble. But to use blazing straw for the purpose is always dangerous in country districts. We have lately heard of a very ingenious and successful mode of entrapping and killing wasps. Place some sugar or strongly sweetened compound on the ground in a garden, and place over it a square hand glass, wedged up an inch or so all round. On this glass, which should have an opening at the apex, lodge another, but a sound one. The wasps, attracted by the sweets, will soon crowd under the lower glass, and when they have well feasted, will ascend into the upper one; there, between the two, they soon become scorched and perish by the heat of the sun shining on the outer glass.

The season of 1864 was most productive for the increase of these prime pests of the apiary, and many hives have severely suffered by their depredations. When once wasps in any number have gained an entrance into a hive, the bees can seldom eject them, and the invaders generally remain until they have freely regaled themselves from the luscious store. They not only consume the honey, but cause a good deal of worry to the legitimate inhabitants of the hive, as well as killing the foremost defenders of it. Wasps being of so much superior strength, it requires, at least, three bees to master one of them.

Having suffered loss in our own apiary from the attacks of wasps, we feel it desirable to give a detailed account of our own troubles from that cause. One of our Ligurian stocks was besieged and worried by wasps to such an extent, that the bees deserted it on the 5th of September, (1864). Fortunately, the bees chose a time for their departure just as we visited the apiary. An unusual turmoil was heard in the hive, such as is experienced at the time of swarming, and on immediately examining the entrance, we observed that the bees were quitting in tumultuous haste. The usual methods that induce bees to settle were tried—amongst others, that of throwing sand up into the air, so as it should fall down amongst the bees on the wing; but they were dispersed in disorder, and their flight extended over three adjacent gardens. We only discovered the clustered bees by diligent search, as the sequel will show. Permission being asked of our next door neighbour, we searched his garden to see if our bees had alighted there; but found that they had passed over. Making a similar application to the owner of the garden adjoining, we entered, having a straw hive in hand, but no bees were there. After looking diligently all round, and climbing the wall, thereby gaining a view of the third garden, we perceived in it unmistakable signs of an unwonted commotion. The occupiers of the house were intently looking at a particular part of the garden, and there was a dustpan and a key, with which the master had been "tanging the bees" to induce them to settle. We quickly made for the proper entrance to the garden, and soon discovered our little wanderers clustered to a large flower vase. Our neighbours, however, were sadly disappointed of their prize, for the gardener had hastily been dispatched into the town to purchase a hive for the welcome colonists. In depriving our neighbours of so unexpected and cheaply acquired a treasure, we could sympathise with their regret, having been much disheartened half-an-hour before at our own loss; but, of course, we could do no other than claim our own bees. We gladly agreed to defray the expense of the straw hive that had been purchased for the sake of our truant swarm. After brushing the bees into the hive, and leaving it propped up with a stick, in order that the stray ones might join, we returned home for an hour or so, to give them time to settle. Judge of our vexation when on returning to fetch the hive home, we found that the refractory creatures had again taken flight, and that all the work was to do over again. The wasps were not to blame for this second flight of the Ligurians; we judged that the swarm had been disturbed by visits from a colony of bees that we discovered were living the life of outlaws under the roof of an adjoining house. Although much disheartened and perplexed, we at once renewed our search, and, upon enquiry, found that the missing bees had taken a southerly course across the turnpike road, and it was therefore necessary to ask permission to search the gardens of the houses opposite. From one of these we observed, on looking through the hedge, that the inhabitants of the next house were on the *qui vive*. On enquiring whether they had seen a colony of bees, the wary old dame replied that she "had no bees but her own," and added that "they were very much excited." Having asked permission to go through the hedge to look at her bees, we soon discovered our Ligurians on the top of the old lady's bee-house. There was no difficulty in identifying our own bees; their yellow rings were as good as a private mark. Quickly hiving the swarm, we took them home, and replaced them in the hive they had quitted. It was almost destitute of honey; but by liberal feeding, and lessening the entrance so that only one bee at a time could find ingress or egress, we succeeded in inducing them to rest in their old home. Thus nearly half a day's exertion was needed to save a fine colony, which would otherwise have been utterly lost by the power of the relentless wasps.

Much watchfulness is needed to prevent the loss of swarms, and and the foregoing incident may serve to suggest the necessity of having hives so located as to be constantly within view, either from the dining room, or of those whose duties oblige them to be near the apiary. If we had not happened to be at hand at the moment this colony started, it would have been irretrievably lost to us. Many swarms and colonies are lost simply because the departure takes place without any one witnessing it. Let us hope that runaway bees may always fall into the hands of those who are as capable of taking care of them as our neighbours appeared to be on the occasion we have described.

Another formidable enemy of bees are the moths. These insects are creatures of the night, as the bees of the day, and they make their way into the hives under cover of darkness, in spite of the bee-sentinels. They deposit their eggs in any crevices in or near the hive that they can find. There the warmth of the hive or of the sheltered situation, causes the eggs speedily to hatch, and then the maggots soon work their way to the comb and larvæ food, which they greedily devour, thereby often bringing about the gradual but certain destruction of the whole community of bees. The best way of keeping moths outside the hives is to lessen the entrance, as before alluded to. Also, in the early spring, the hives should be lifted from their floorboards, which must then be made thoroughly clean, and all crevices and corners about the hive and stand should be scraped, so as to get rid of all eggs of moths and other insects before the warm weather hatches them or enables them to do mischief. The bee-moth is not so troublesome in England as it is in America and some parts of Germany; but still its encroachments should be carefully guarded against in this country, for if not, it may easily increase to a very serious extent.

## DRAINING HONEY FROM THE COMBS.

Those of our readers who prefer eating " run honey" to honey in the comb, may be glad of some instruction as to the best way of separating the two. For this purpose, it is better to let the honey run without squeezing, in order to preserve both its transparency and flavour.

Take a sharp knife, and slice the combs on both sides, keeping the knife parallel with the partition wall, so that every cell may be laid open. Place these broken combs in a sieve, or on a piece of muslin stretched across and tied round the opening of a pan or large mouthed jar. Allow the honey to flow out of the combs spontaneously, and reserve the squeezing process for a separate jar, so that the honey of the first drained jar may be perfectly pure, both in appearance and flavour. That which has pressure put on it will be waxy in flavour and thick. Some persons recommend that the opened combs be placed in the sun, as the heat will cause the honey to run more freely. The great disadvantage of this is, the temptation the honey will offer to bees, who will be eager to gain a share. Honey whilst in the combs keeps remarkably well when left in the supers; if cut out, the combs should be folded in writing paper and sealed up, so as effectually to prevent free entrance of air; they should then be placed in a warm dry closet.

Honey, like most vegetable products, should be fresh every year. It may easily be kept from one season to another; but when kept beyond that time, unless very carefully stored in a warm temperature, it will crystalize in the comb, and it is liable to ferment when in jars separated from the comb.

## GENERAL REMARKS.

Every bee-keeper should be a book-keeper; that is, so far as to have a permanent record of the events of the apiary and the fortunes of his bees. A book similar to a tradesman's journal would be very suitable for the purpose. In it he should note down the date of the first swarm of the season especially, and those of the other swarms also; and in autumn, the quantity of honey taken from each hive should be entered, with remarks on the probable size of the various stocks. These particulars will not only be interesting for the bee-keeper to turn to in winter, but will be of practical service in enabling him to know the exact age and probable strength of each stock. The bee-book may also be contrived to show the total amount of honey that the bees have produced for their owner, and the net money profit of the apiary. A simple and clear account like this—provided, by the bye, that it does show a satisfactory balance—will be very useful for inducing cottagers and farm labourers to start bee-keeping. Nothing like ocular demonstration for this class. The "humane" apiarian will reason with them in vain until he shows them a monster "skep" of honey, and mentions the price that it will fetch in the market. When convinced that the depriving system will pay, the cottager will gladly adopt it.

A writer in the *Quarterly Review* gives the following good advice:—"Don't bore the cottager with long lectures; don't heap upon him many little books; but give him a hive of the best construction, show him the management, and then *buy his honey;* buy all he brings, even though you should have to give the surplus to some gardenless widow. But only buy such as comes from an improved hive—and you cannot easily be deceived in this,—one which preserves the bees and betters the honey. Then, *when you pay him,* you may read to him, if you will, the wise rules of old Butler," *exempli gratiâ:*—

"If thou wilt have the favour of thy bees that they sting thee not, thou must not be unchaste or uncleanly; thou must not come among them with a stinking breath, caused either through eating of leeks, onions, or garlic, or by any other means, the noisomeness whereof is corrected by a cup of beer; thou must not be given to surfeiting or drunkenness; thou must not come puffing or blowing unto them, neither hastily stir among them, nor violently defend thyself when they seem to threaten thee; but, softly moving by, thy hand before thy face, gently put them by; and, lastly, thou must be no stranger to them. In a word, (or rather, in five words,) be chaste, sweet, sober, quiet, familiar; so they will love thee and know thee from all others."

Allusion having been made to the profit that may be gained by the judicious management of bees, we will illustrate that point by relating an anecdote of a certain French *curé.** It is one which

---

* This story, in a disguised form, or—as the writer would say—an improved form, was quoted in the *Cornhill Magazine* some time ago. In transforming the bee-keeping *curé* into an English clergyman the effect was cleverly

may be suggestive to some of the rural clergy in this country, who might almost as easily keep an apiary as they do a garden or an orchard.

A good French bishop, in paying his annual visit to his clergy, was very much afflicted by the representations they made to him of their extreme poverty, which indeed the appearance of their houses and families corroborated. Deploring the sad state of things which had reduced them to such a condition, he arrived at the house of a curate who, living amongst a poorer set of parishioners than any he had yet visited, would, he feared, be in a still more woful plight than the rest. Contrary, however, to his expectations, he found the appearance of this remote parsonage to be superior to those he had already visited. Everything about the house wore the aspect of comfort and plenty. The good bishop was amazed. " How is this, my friend," said he, " you are the first pastor I have met with having a cheerful face and a plentiful board ! Have you any income independent of your cure ? "  " Yes, Sire," said the pastor, " I have : my family would starve on the pittance I receive from the poor people that I instruct. If you will walk into the garden, I will show you the stock that yields me such excellent interest." On going into the garden, he showed the bishop a long range of bee-hives.  " There," said he, " is the bank from which I draw an annual dividend, and it is one that never stops payment." His harvest of honey enabled him almost to dispense with the use of sugar, leaving him a considerable quantity of it for disposal in the market ; of the coarser portions he made a tolerable substitute for wine, and the sale of his wax nearly paid his shoemaker's bill. Ever afterwards, when any of the clergy complained to the bishop of poverty, he would say to them " Keep bees ! keep bees ! " In this succinct advice, extending it to laity as well as clergy in rural districts, we heartily join, believing that in this country a ten times greater number of hives might be successfully kept than are now established.  In a very practical sense, the oft repeated lines of Gray are strictly true :—

> " Full many a flower is born to blush unseen,
> And *waste* its fragrance on the desert air."

enhanced, especially as to the dismay of the decorous English prelate in hearing that his poor brother in the Church had turned " manufacturer;" but then the *vraisemblance* of the story, as we have it, was destroyed.

An apiary in the garden of every village clergyman would afford the means of economising this unclaimed bounty of Providence.

Bees may be very inexpensively and profitably kept in the Cottager's hive (see page 34), which will be found a very productive one. It is true that it has not the appliances of windows and bell glasses; for the cottager is not supposed so much to care for his hives as a source of amusement; his object in bee-keeping is simply the profit it may bring. For those of our readers who wish to have united the facility of observing the bees with that of the plentiful production of honey, we would especially recommend the "Improved Cottage" hive, described at page 28. If inclined to go to a little further expense, the hives numbered 1, 2, 3, and 7, all afford constant opportunity for inspection of the bees, and allow of their working freely in the most natural manner.

There are few hobbies which cost so little outlay as the keeping of bees. Once the "plant" of hives is purchased, there is little, if any, additional expense, and always a probability of a fair return. If honey be obtainable, the bees will find it; they work for nothing, and provide themselves with sustenance, requiring only a very little labour from their keepers, and that labour is of a pleasing and instructive kind.

To the advanced and skilful apiarian we would especially commend the use of the Bar-and-frame hives. With these, as we have attempted to show, the bee-keeper has a full command over his hives and bees. Many mistakes, it is true, have been made by uninitiated bee-keepers in using the more elaborate hives. Being struck with the remarkable facilities afforded by these superior hives for the extraction of any one comb, and, perhaps, fascinated with their easy sway over so highly organized a community, these new-fangled bee-keepers have acquired a habit of perpetually and incautiously meddling with the bees. The inevitable results in such cases are distress to the bees, impoverishment of the stocks, and loss and vexation to the over zealous apiarian. All these things may be avoided, if it be remembered that there are first steps in bee-keeping, as well as in croquet, chemistry, or cricket. In bee-keeping, as in floriculture, it is a great point to know when to "let well alone." There is no florist, however anxious for a prize, who would be continually pulling up his plants to see how their roots were growing. Doubtless, the full control which the bars and frames afford over the inmost

I

recesses of the hives, is a great temptation to the bee-keeper; but, if he yields too readily to it, he will imperil his chance of profit, and deprive himself of that continuous source of interest, which a judicious apiarian always enjoys.

Many persons who are well informed on most subjects, are extraordinarily ignorant of the natural history of bees, and the economy of the bee-hive. Perhaps we might venture to suggest that, more pains should be taken at schools or by parents to inform young persons on this, in connection with kindred subjects. As an amusing illustration of the ignorance referred to, we transcribe an order we received a short time since from a seminary in the north of England. The young gentleman thus writes:—"Master —— presents his compliments to Messrs. Neighbour, and begs they will send him a swarm of bees; he encloses *six postage stamps*, and hopes they will send him a *good* swarm." This embryo naturalist was evidently of a mercantile turn, and had a mind to buy in the cheapest market, for in a postscript he adds:—"Please let it be fourpence, if you can!" We need scarcely say that in reply we endeavoured to enlighten our juvenile correspondent as to what constituted a swarm of bees, and returned the stamps, with our thanks.

The culture of bees would be greatly promoted, if a knowledge of it were considered necessary as one of the regular qualifications of a gardener. So little time is needed to gain the skill requisite for the tendance of an apiary, that it seems only reasonable to expect it of a well taught gardener, and he should feel a pleasure in the circumstance of its forming a part of his duties. In Germany, where a country gentleman's table is kept constantly supplied with fresh honey, the gardeners are expected to understand the management of hives; and in Bavaria, modern bee culture is taught in the colleges to all the horticultural students. Travellers in Switzerland will call to mind the almost invariable practice of placing new honey on the breakfast tables at hotels in that country.

Some writers on bee-culture attach much importance to the particular position in which an apiary stands, and the aspect towards which it faces. A southern, or rather a south-eastern aspect is the one which we have already recommended. Our reason for this preference is, that we deem it very important for the bees to have the first of the morning sun. Bees are early risers, and should have every inducement given them for the maintenance of so excellent a practice.    A few years since, many strong opinions were expressed in favour of a northern aspect for hives.    The chief reason given for those opinions, though very plausible, appears to us to be a very partial and inadequate one.    It was said that, when the hives face the south, the bees may, like the incautious swallow in the fable, be tempted to fly abroad in the transient winter sunshine, and then perish in the freezing atmosphere when a passing cloud intervenes.    But it is a very easy matter, if considered needful, to screen the entrance by fixing up matting so as to inter-cept the rays of the sun.    At our own apiary we make no alteration in winter, under the belief that the bees will take care of themselves, and they seldom venture out when the weather is unsuitable.

With hives exposed in the open garden, it is a good practice to wind hay-bands round them in frosty weather, as such a protection enables the bees to resist the cold.

When a thaw occurs, everything, both in and out of doors, has a great deal of dampness about it.    The combs of a hive are not exempt from this, so that it is advisable to have slight upward ventilation in winter.    Holes the size of a pin's head allow of the escape of a good deal of bad air, which is generated by the exhalations of the bees, as well as by the dampness before referred to.    These holes being small, do not create sufficient draft through the hives to be pernicious; if closed up by propolis, are readily reopened with a pin.    With wooden hives in winter, a bell glass is often found to be useful; it should be placed over the hole in the crown-board, with a zinc trough to receive the condensed moisture.

In summer bees do much towards ventilating their own stock-hives.    The observant apiarian will not fail to remark how, on a warm day, several of the little creatures will stand at the entrance with their abdomens slightly raised, and their twinkling wings in rapid motion, producing a current of air inwards; while another set are engaged in like manner, driving the bad air out, so that a supply of pure oxygen is conveyed to the crowded inmates.    In this fanning operation their wings vibrate with such rapidity, that their shape is as indistinct as are the spokes of a wheel when revolving in rapid centrifugal motion.

This important office entails great physical exertion on the part of the bees, and they relieve each other in detachments.

Some bee-keepers find an adapting board convenient for placing underneath straw supers, as it facilitates their removal. These boards are made of mahogany half an inch thick, with a hole in the centre corresponding with that in the stock hive. We do not consider it necessary to fix cross sticks in the straw stock hives, as is frequently done; but if the apiarian prefers to have his hives so furnished, there is no serious objection to it. These observations refer to our Cottager's hive (page 34).

There is another little matter of detail that should be named here; that is, the necessity of the bee-keeper always having a common hive in readiness near the bees, so as to be able to secure any swarm which may unexpectedly start.

Here our pleasant task must close. We trust that all information has been given that is needful to enable the practical bee-keeper to begin business, and the scientific apiarian to commence his observations. By way of illustrating the two characters combined, we will conclude by quoting another simple idyl by the German bee-keeper, Herr Braun, whose winter musings we have already presented to the reader.

[*From " The Journal of Horticulture."*]

## ON THE FIRST FLIGHT OF BEES IN SPRING.

*By* ADALBERT BRAUN, *Translated by* "A DEVONSHIRE BEE-KEEPER."

Hark ! what is so gaily humming
    In the little garden there ?
Hark ! what is so briskly whizzing
    Through the still and silent air ?

Friend, it is our bees—the darlings—
    Now enliven'd by the Spring;
Yes, the winter is departed,
    And once more they're on the wing.

Happy he, who winter's perils
    All his stocks brings safely through;
Thank Him, of all good the Giver—
    Faithful Watchman He, and true.

Of my own are none departed,
    All as yet unhurt remain;
Though no longer rich in honey,
    Yet is Spring returned again !

Come, and let us view them nearer—
    Enter by the garden gate ;—
So—stand still, and watch their doings—
    Light your pipe, and patient wait.

See how busily they traverse
    To their pasturage and back,
That they may by toil unwearied
    Save the commonwealth from wrack.

Look, O look! what loads of pollen,
    Bring they in with heedful care.
Nurslings, fear not; for your cravings
    Here's sufficient and to spare.

How they dart and how they hurtle
    Through the genial balmy air!
To the mountains—to the meadows—
    'Tis the scent attracts them there!

There they dexterously rifle
    Nectar from each flow'r in bloom;
Toil they for our honey harvest,
    For us fill the honey-room.

Yes, our bees, our precious darlings,
    We salute you all to-day;
For your life is our enjoyment—
    Winter's sleep has pass'd away.

Grant prosperity, O Heaven!
    To the new-born honey-year—
Give thy favour—give thy blessing—
    To these objects of our care.

Now let each attentive guardian
    In devoted service strive
For the proud, the Matron-monarch—
    Sov'reign of the honey-hive.

So that we may learn by watching
    Who that in the noon-tide glance,
Or in midnight's darkest moments,
    Summons her to Hymen's dance.*

Ev'ry bee-hive calls for patience,
    Whilst great HALLER's lessons teach
Without patience Nature's secrets
    None successfully can reach.

        T. W. WOODBURY, *Mount Radford, Exeter.*

---

   * This point cannot now be considered doubtful, but it must be remembered that Herr Braun's verses were written eighteen years ago.

In conclusion, we would remind all bee-keepers who earnestly desire success, and who hope to draw pecuniary profit from their pursuit, of the golden rule in bee-keeping :—" Keep your stocks strong." In exercising the assiduous attention and persevering effort, which that maxim enjoins, they will not only be regarded as *bee-keepers*, but, as Mr. Langstroth says, will acquire a right to the title of *bee-masters*.

# APPENDIX.

---

## TESTIMONIALS OF THE PRESS.

### GREAT EXHIBITION 1851.

The " Working Apiary" in the Great Exhibition of 1851, will long live in remembrance of the many thousand visitors who witnessed with much interest the matchless industry of its busy occupants.

We extract the following from many notices that appeared in the public journals relative thereto.

In noticing the hives exhibited in the Crystal Palace, I would say, first and foremost in my opinion stands Mr. TAYLOR's Eight-bar Hive, and Messrs. NEIGHBOUR AND SON's Improved Cottage Hive, both exhibited by Messrs. NEIGHBOUR.—*J. H. Payne, see Cottage Gardener, Nos.* 169, 170.

### From the " Illustrated London News."

Messrs. NEIGHBOUR's Apiary consists of a large glass case, with parts of the sides covered with perforated zinc, for the sake of ventilation. This apiary contains three hives; first NEIGHBOUR's Ventilating Box-Hive, containing from 15,000 to 20,000 bees, which were hived on the 30th of April of the present year, the day before that of the opening of the Great Exhibition; NEIGHBOUR's Observatory Glass Hive, containing about the same number as the box-hive; and a two storied square box-hive, with sloping roof. From this latter, however, the bees decamped within a week after they had been hived, owing to some disturbance, or perhaps, to the dislike taken by the bees to their new habitation. The Ventilating Box-Hive is, in shape, square, having windows and shutters. The entrance is at the back, enabling the bees to go to Kensington Gardens, or other resorts, when they please. Above the wooden box is placed a bell glass, into which the bees ascend to work through circular opening in the top of the square box. In the top of the bell-glass is an aperture through which is inserted a tubular trunk of perforated zinc, to take off the moisture from within. The Observatory Hive is of glass, with a superior crystal compartment, an opening being formed between the two ; the bees are at present forming a comb in this upper glass, which affords a very interesting sight, as generally speaking, the bees are in such a cluster when at work that one can scarcely view their mathematically formed cells.

.

A straw cover is suspended over the upper compartment by a rope over a pulley, which cover is raised up by the attendant at pleasure. The larger or bottom compartment rests on a wooden floor, which has a circular groove sinking therein to receive the bell-glass. A landing-place projecting, as usual, with sunken way, to enable the bees to pass in and out of their habitation, completes this contrivance.

In addition to Mr. NEIGHBOUR's Crystal Apiary, he also exhibits a Cottager's Straw Hive, TAYLOR's Amateur Bee-Hive, a Glass Hive, NUTT's Patent Collateral Hive, the Ladies' Observatory Hive, NEIGHBOUR's Improved Cottage Hive, and PAYNE's Cottage Hive.

The Cottager's Hive is simply that of the form we find in use in most parts of the country, where the industrious cottagers or their wives, by a little attention to their interesting little labourers, are enabled to add something to their usually scanty earnings. This kind of hive is usually made of straw, resting on a circular wooden board, with part of the board or floor projecting in front as a landing place for the bees, which enter under the edge of the straw by means of a sinking in the floor.

TAYLOR's Amateur's Bee-Hive consists of three small square boxes, one above another, with a roof over the top story; the ventilation being effected by perforations under the eaves; each side of every story has a window and shutter. The landing place is in front of the bottom story, and the entrance to the hive is a long slit about ⅜ inch high.

The Glass Hive or Ladies' Observatory Hive, is similar to that in which the bees are at work in Mr. NEIGHBOUR's Apiary already mentioned, but on account of the number of bees at work therein, and the extent of comb already effected, the interior perches cannot be seen. These wooden perches are arranged in parallel lines, leaving a space next the glass all round, the whole being framed together with a bar at right angles, and resting on an upright support in the middle.

The Improved Cottage Hive of the same exhibitor consists of a straw circular lower compartment, having windows and outside shutters. A thermometer is placed just inside one of the windows. The floor is of wood, with a landing place and sunken way, as already mentioned in some of the other hives. In the top, which is also of wood, are three circular perforations, each of about two inches in diameter; above which are placed as many bell-glasses. There is a small hole in the top of each of the glasses, through which a perforated tubular trunk is inserted, for the sake of taking off the moisture from the interior of the hive. Within the glass is a feeding-trough of zinc, circular in shape, with a floating perforated floor, on which the bees alight, and in the the winter season regale themselves with the honey which is found in the various perforations, as it floats up to the level of the honey, contained in the small filling-trough, through which the honey, or beer and sugar, is poured. The glasses are covered with a straw cap, removable at pleasure.

Messrs. NEIGHBOUR's contributions are completed with, tin perforated umigators, by the use of which the bees are stupefied for a while, when required to be moved from one hive to another; and specimens of honey and honeycomb of the season 1850.

*From the " Express."*

BEES AND BEE-HIVES.—In the North-East Gallery directly under the Transept are arranged by Messrs. NEIGHBOUR, of Holborn, several descriptions of bee-hives, which it will be interesting to many of our readers to examine, as this branch of rural economy is claiming much general and deserved attention throughout the country. The novelty of these hives consists in the facilities that are afforded in taking therefrom at any time of the gathering season the purest honey without destroying or even injuring the bees, thus humanely superseding the barbarous and hateful system of murdering these interesting insects, to obtain the produce of their industry.

Immediately adjoining the group of untenanted bee-hives may be observed living hives with the bees most industriously at work. These useful little creatures have been highly honoured by the Executive Committee, for of all the animal workers that contribute to the interest of the Exhibition they alone are allowed therein to display their matchless ingenuity and skill. By a simple contrivance the bees are allowed egress and ingress without in the least degree molesting the visitors, thus enabling the admirers of the works of nature to view the whole process of forming the cells and depositing the honey therein.

Within these few days Messrs. NEIGHBOUR have added to the Apiary, a bee-hive constructed entirely of glass, protected by a cover neatly made of straw, but so contrived, that on application to the attendant can be removed instantly, thus illustrating more particularly the curious workmanship of these amusing insects.

Her Majesty the Queen, and the Prince Consort, with the Royal Children, were some time engaged in watching with deep interest the busy scene before them, and putting many questions relating to the habits and economy of the honey bee.

## INTERNATIONAL EXHIBITION 1861.

*From the " Illustrated London News," August* 16, 1862.

One of the most interesting and instructive objects in the Exhibition, is a transparent hive, in which the bees may be seen at full work. Among the collection of bee-hives exhibited by Messrs. NEIGHBOUR AND SON, is one of glass, stocked with a colony of Italian Alp bees. Here the queen bee may be seen surrounded by her subjects, which pay the most deferential attention to their sovereign. Through an aperture cut in the wall, the busy throng of bees are continually passing and repassing. They go out at their pleasure into the open court, fly over the annexe into the grounds of the Horticultural Society and other adjacent gardens, and return laden with sweets.

*From the " Journal of Horticulture," October* 21, 1862.

NEIGHBOUR, G. & SONS, 149, Regent Street, and 127, Holborn, No. 2157, have a very handsome and complete stall, on ascending the steps of which we found a flourishing stock of Ligurians, apparently not at all ashamed of the public position which they occupied, and working vigorously in the full light of day. The queen, one of the largest and finest-coloured we have met with, was perambulating the combs and receiving the homage of her subjects, stopping frequently to deposit an egg in every empty cell. The hive itself was a "Woodbury Unicomb," handsomely got up in mahogany, invented as its name implies, by our valued correspondent. " A DEVONSHIRE BEE-KEEPER," the construction of which will be most readily understood by an inspection of the engraving at page 46. Its distinctive features are, the adaptation of the moveable-bar system to unicomb-hives, by which any colony in an apiary of " Woodbury hives" can be placed in the unicomb-hive in a few minutes, and the use of "outside venetians," or "sun blinds." as they are called, instead of the usual impervious shutters. By this contrivance light is never excluded, so that when the hive is open for inspection, all its inmates continue their avocation with their accustomed regularity, and a quiet and orderly scene is presented to the spectator, instead of the hubbub and confusion which ensues in ordinary unicomb-hives. On the left-hand side of the unicomb hangs a beautifully executed drawing of a Ligurian queen bee magnified, together with the queen worker and drone of *Apis Ligustica*, of the natural size. Immediately under the drawing is placed a square glass super containing nearly 40lbs. of the finest honeycomb. On the right of the unicomb-hive is another super of the same description, containing nearly 30lbs. of the purest honey. These supers are, undoubtedly, by far the finest in the Exhibition, and are the first worked in England by Ligurian bees, being from the apiary of " A DEVONSHIRE BEE-KEEPER." In addition to these the most striking objects, are shewn Neighbour's Improved Single Box and Cottage Hives, Taylor's Bar-Hives, Woodbury Frame and Bar-hives, the new Bottle-feeder, and bee apparatus of every description. It will be apparent from the foregoing, that Messrs. Neighbour's stall is well worth inspection, although the various novelties it contains appear to have met with but scant appreciation by the Jury, who merely awarded to them that "honourable mention" so lavishly accorded to the far less deserving objects.

*From the " Illustrated News of the World," September 6, 1862.*

One of the most interesting and instructive objects is the honey bee at full work in transparent hives. In the International Exhibition, Class 9, Eastern Annexe, Messrs. NEIGHBOUR and Son, of Holborn and Regent-street, exhibit, amongst a collection of the most approved beehives and apparatus, a glass hive, stocked with a colony of Italian Alp bees; the hive is so constructed as to admit of easily seeing the queen, surrounded by the working bees. Contrary to the long established notion that the bees work only in the dark this hive is completely open to broad daylight. The bees do not manifest the least dislike to the exposure, and they are not discomfited when light is occasionally admitted for inspecting them. It is obvious that a knowledge of this new feature must tend to a more general acquaintance with the habits and hidden mysteries of the bee than has hitherto been the case. The queen may be seen depositing the eggs in the cells; in this manner she goes on multiplying the species, the working bees surrounding her, and paying the most deferential attention, with their heads always towards her. Not the least interesting part is to watch the entrance; facility is afforded for doing so, the sunken way communicating with the hive being covered with a flat piece of glass; the busy throng, pass and repass through the apperture cut in the wall, so that the bees go out at their pleasure into the open court, fly over the Annexe into the Horticultural and other adjacent gardens, and return laden with crystal sweets gathered from the flowers. The novelty of being able to inspect living bees, and those of a new variety, as easily as goods in a shop window, will well repay the trouble of finding Messrs. Neighbour's stand. These gentlemen will no doubt cheerfully give any information that may be required.

*From the "Gardener's Weekly Magazine," September 1, 1862. Conducted by Shirley Hibberd, Esq., F.R.H.S.*

Neighbour and Son, 149, Regent Street, London, (2157).—This is the most important of the exhibitions in this department. The "Bees at work" are in hives open to the inspection of visitors, the bees passing out through tubes to the open air, and not being visible within the building except through the glass of the hives. The collection of hives of all kinds is complete and interesting, and we subjoin a figure of the stand (see page 100) to show how bees as well as hives may be exhibited conveniently. Amongst the various contrivances exhibited by Messrs. Neighbour, Nutt's Collateral Hive has an important place, and though very fancifully got up, and therefore very attractive to amateur bee-keepers, we must make the same objection to it as we have above to other forms of the same from different makers. The Single-box Hive, the Taylor's Shallow Eight-bar Hive, are the best bee-boxes in this collection, and every way admirable. Here are no fancy ventilators which the bees will close up, nor provoking side boxes which they will hesitate to enter, and from which it will be hard to dislodge them in order to get them to winter in the "pavilion." Whoever begins bee-keeping with either of these will have a fair chance of success. The most popular of the hives is that called the "Improved Cottage." Its popularity no doubt is due to the compromise between wood and straw which it accomplishes. People cannot get rid of the idea that a beehive *must* be made of straw, though it is a material so ill adapted for union of swarms, supering and other operations of advanced bee culture. The "Cottager's Hive" is well adapted for "those apiarians who are desirous of setting their poorer neighbours in the way of keeping bees on the improved system." It consists of stock hive, small super-hive, and straw cover, and is on the principle of Payne's, which has been most successful among country people who have got so far as to prefer keeping, to killing their bees. The "Woodbury Bar and Frame Hive" is a novel construction, combining all the best features of the best bar boxes, and adding some new ones of great value and importance. We recommend every bee-keeper to become possessed of this admirable contrivance, with which Mr. Woodbury has accomplished wonderful things in the multiplication of the new race of Ligurian bees. In general form and proportions it resembles Taylor's and Tegetmeier's boxes, but in the arrangement of the bars it is unique. The stock box is furnished with ten moveable bars and frames, after the German plan. Each bar has a projection running along the under side; this ridge is chosen by the bees for the foundation of combs, rendering guide combs unnecessary. The supers have glass sides and eight bars, so that the operator need never be in doubt when to add another box above or take away the harvest.

The "Unicomb," or one-comb observatory-hive, is intended solely for purposes of observation. and though furnished with doors, to keep up a uniform degree of heat, Messrs. Neighbour have found in their experience at the Regent's Park Gardens and elsewhere, that the bees manifest no dislike to a continual exposure to light. As this elegant contrivance can be placed in the window of a drawing-room, it is adapted to the amusement and instruction of the family circle. as well as to the more serious objects of the etomologist and scientific apiarian. All that is necessary is to connect the outlet with the open air by means of a length of tubing or wooden tunnel, and the bees pass in and out without obtaining access to the room, and all the mysteries of the hive are open to daily observation. There are numerous other hives, bee-feeders, bee armour, &c., &c., which we have not space to notice, but which we advise our apiarian friends to inspect, as the collection of Messrs. Neighbour illustrates fully every department of this interesting subject.

## BATH AND WEST OF ENGLAND AGRICULTURAL SHOW AT EXETER, IN JUNE, 1863.

*From the " Journal of Horticulture," June 23, 1863.*

A novel feature in the Exhibition of the Bath and West of England Agricultural Society which took place at Exeter last week, was the stall of Messrs. NEIGHBOUR & SONS, in which were exhibited bees at work in glass hives, and apiarian appliances of every description. There were two Ligurian stocks of bees at full work, one in a full sized Woodbury Unicomb Hive, having been brought from London for the occasion, and the other in a smaller hive of the same description being from the neighbouring apiary of our valued correspondent, "A DEVONSHIRE BEE-KEEPER." Amongst the hives exhibited, the Woodbury Frame Hive in straw appeared both novel and good, whilst amongst the apparatus, artificial combs and the stereotyped plates for making them, seemed to us the most worthy of attention.

There was a remarkably curious specimen of artificial combs or partition wall partially fabricated into complete comb by the bees, which struck us as being well worth examination, showing, as it did, the various stages by which this transformation is effected, and being calculated to throw light on the problem as to the mode in which bees construct their combs. It is almost unnecessary to state, that this unique and instructive stall was crowded throughout the week, and we hope its financial results were such as will lead Messrs. NEIGHBOUR to continue their attendance at the Society's meetings.

*From the " Western Times," Exeter, June 12, 1863.*

FOR THE LITTLE BUSY BEE.—Next to the poultry tents, and set back against the yard fencing, is the exhibition of Mr. GEORGE NEIGHBOUR & SONS, 127, High Holborn, and 149, Regent Street, London. inventors and manufacturers of improved bee-hives for taking honey without the destruction of the bees. The savage knows where to find the nest of the wild bee, and how to get at his honey; but all the improvement upon the covetousness of the savage made by the long after ages of the world to modern times, was to find means of luring the pattern of industry to a convenient atelier where he might be more easily, first murdered and then robbed. Their habits early attracted the attention of some of the best observers of ancient as well as modern times; Cicero and Pliny tell of the philosopher Hyliscus quitting human society and retiring to the desert to contemplate their peaceful industry. The ancient poet in his *Sic vos non vobis* plaintively sings over bee and beast, living, or rather dying, not for themselves, but the lord of creation, yet was it left to modern times—very modern times—to join the sentiment of humanity to the rapacity of the barbarian. Mr. NEIGHBOUR has a very complete collection of specimens of the ingenious and successful contrivances in the construction of hives for the double object of preserving the honey and the life of the bee; and also, subsidiarily of promoting its comfort during its busy and useful life. We are not allowed to forget here, that we have residing in our city one of the first apiarians in the kingdom—Mr. THOMAS WOODBURY, of Mount Radford. If the bee philosophy be his hobby, we may recollect that all great discoveries and improvements owe their existence to men who have had the power and the will to concentrate their faculties upon a single object. One proof of his genius in this his favourite department of action, is seen among this collection of Mr. NEIGHBOUR's in the "WOODBURY Unicomb Hive." It might be when closed up, for ought that appears, a neat case of books; but on opening two doors of the Venetian blind pattern, back and front, we see between the glass walls, the insect city exposed to view with all the population in action. There it may be seen

> How skilfully she builds her cell;
> How neat she spreads her wax;
> And labours hard to store it well
> With the sweet food she makes.

Some of the hives are constructed chiefly for the purpose of promoting a philosophic observation of the bee's habits and methods of procedure in his wonderful work. "NEIGHBOUR's Unicomb Observatory Hive" is a great novelty, being constructed with glass sides, the hidden mysteries of the hive being exposed to the full light of day. "HUBER's Book or Leaf Hive" is constructed to facilitate the object of the scientific apiarian. But the class of hives which will most interest those desirous of promoting bee-keeping among the many will be those for the cottage. There can be no doubt that many a poor cottager in the country, if he could be made to see the advantage it would be to him, and were taught the most economical and successful way of managing this species of "live stock," would add thereby something considerable to his small earnings in the course of a summer. Members of Cottage Garden Societies have turned their attention to it very generally; but to get the thing well afloat, requires in every district the devotion of some earnest enthusiast who will take up the apostolic rule of action "This one thing I do." There is the No. 5, "Improved Cottage Hive," in which three bell glasses are employed, enabling you to take a glass of the purest honey from the hive in the most vigorous period of the season. Then there are other hives of simpler construction and less expensive, but all illustrative of the sentiment of humanity which seeks to preserve from wanton destruction those useful and interesting auxiliaries to our luxury and comfort. This comparatively unimportant stand, in point of size, cannot but attract the attention of a large number of visitors, especially of the ladies and the clergy, who are desirous of promoting the cultivation of the bee among the poor. To heighten the interest for the curious, in one of Mr. Woodbury's hives the bees are all alive and at it, and for those who are disposed to go further into the subject, information is available touching this fashionable, profitable, and domesticated member of the Apiarian family, the Ligurian and Italian Alp-bee. Our old dark-coated delver is threatened with supercession, just as the black rat was driven off by the Norwegian invader, now in possession, and as the old races of cattle are being metamorphosed into the sleck, shapely, beef-bearing, small-boned animals of the present time.

*From the " Devon Weekly Times," June 8, 1863.*

Bees.—Messrs. Neighbour & Sons, of London, are exhibitors of two Woodbury Unicomb Hives, showing the royal and common bees in full work. These hives are very ingeniously constructed, and were invented by Mr. T. Woodbury, of Mount Radford. Among other apiarian attractions, we may mention the improved Cottage and Cottager's Hives, which are well worthy the notice of those for whom they are designed, and the Ladies' Observatory Hive. The Messrs. Neighbour also exhibit Ligurian bees.

*From " Woolmer's Exeter Gazette," June 12, 1863.*

Improved Bee-Hives.—At a stand near the poultry tents, are exhibited Neighbour's Improved Bee-hives for the taking of honey without the destruction of bees. The hives are stocked with the famous Ligurian bee. The Unicomb Observatory Hive is constructed with glass sides, so that the whole of the movements of the Apiarian colony are visible, including the proceedings of the queen and her court. This and some of the other descriptions of hives manufactured by Messrs. Neighbour are invented by T. W. Woodbury, Esq. They are furnished with moveable bars, after the German fashion. Each bar has a projection along the underside; this ridge being waxed, induces the bees to build parallel combs—thus obviating the necessity for a guide comb. This description of hive is best suited for the Ligurian or Italian Alp-bee. Stocks of this species, now so much in repute, may be obtained of Messrs. Neighbour 149, Regent Street, London.

## ROYAL AGRICULTURAL SHOW, NEWCASTLE, 1864.

*From the " Northern Daily Express," July 22,—(Published at Newcastle.)*

A Model Factory.—Stand 194—G. Neighbour & Sons. Regent Street, and High Holborn, London. We have heard of model farms and model lodging houses for the working classes, but it was reserved for the Royal Agricultural Society's Meeting in 1864 to introduce to our notice a model factory, where we may see representatives of the working classes busily engaged in their daily avocations. The stand which we have quoted above, may afford fruitful study to such philanthropists as the Earl of Shaftesbury, who make it their benevolent aim to elevate the masses, and the lesson here given from actual life will not be lost upon working men themselves. There are several striking features worthy of notice in the "model factory." We can clearly perceive that it has been established on a principle which is essential to the success of any great concern—namely : the principle of a good understanding amongst the operatives themselves, and between them and the head of the establishment. What strikes us in this model factory is the unity of action which reigns throughout. There is no jostling of rival interests, and no misunderstandings, or cross purposes. The operatives in this establishment are so numerous that we question if any one has as yet been able to count their number, and yet all seem to be working in perfect harmony, their joint labour continually leading to one beautiful and sublime result. Another feature specially noticeable in the establishment in question is the principle of subordination. Singular to say, that while the operatives are males* the foreman of this model factory is a female; but that circumstance need not shock the sensibilities of our fair friends any more than it ought to offend the prejudices of the sterner sex, inasmuch as the mighty empire of Great Britain is ruled by the gentle hand of a female; and moreover, in the one case as in the other, the presiding genius, amids all her official cares and duties, takes care to preserve the modesty of her sex. She never in the slightest degree obtrudes herself needlessly on public observation, and probably on that very account the respect shown to her by her subjects is the more profound and devoted. There is, however, one particular in which we would take leave to demur to the idea of this factory being in every respect regarded as a " model." We have not been able to discover that there is any particular period of the day in which the operatives are allowed to take refreshments. We, in England, have been accustomed to regard the dinner hour somewhat in the light of a sacred institution. And if the council of the Royal Agricultural Society mean to set this up as a model institution, we are of opinion that some explanation on this point is desirable. Indeed we have not been able to discover that the operatives in this establishment take any refreshment whatever. If they do, it must be " on the sly," vulgarly speaking. There is one peculiarity, however, which must tend to popularize this institution, and which has served to make it one of the most attractive objects on the show ground. It is the fashion in all the great factories which abound on the banks of the Tyne and throughout the country generally to act on the principle of exclusiveness to a very great extent, and perhaps wisely so. As you approach the door you see an intimation in legible characters, "No admission except on business." This may be very proper, but it is rather tantalising. In the model factory which we are now describing all the operations are open to inspection. Every action is patent to the eye of the spectator. This has been effected by a skilful contrivance, and it is this contrivance in fact, which has entitled the inventor to obtain a place in the show ground for his model factory, which he describes by the somewhat ambiguous term of " a new implement." But our readers may wish to learn what

* The reporter was in error as to the sex of the workers.

is the staple manufacture of this wonderful workshop. We reply—"honey."
The factory we speak of is nothing more nor less than a beehive; or, to quote
from the catalogue, "An Unicomb Observatory Bee-hive," with living Italian
Alpine bees at full work; it was invented by T. W. WOODBURY, of Exeter;
and is improved and manufactured by the exhibitors. As implied by its
name this hive has one comb, so that both sides are fully exposed to the light
of day, thus allowing of an easy inspection of the queen-bee, surrounded
by her retinue.

# INDEX.

K

# NEIGHBOUR'S
# IMPROVED BEE-HIVES,

FOR '

## TAKING HONEY WITHOUT THE DESTRUCTION OF THE BEES.

DRAWINGS AND DETAILED LISTS FORWARDED ON RECEIPT OF TWO POSTAGE STAMPS.

———o◦|◦⇒|◦o◦———

1.  Nutt's Collateral Bee-Hive  -       -       -       -  6  15  0
    Stand for ditto, 16s.

2.  Neighbour's Improved Single-box Hive  -       -    3   3  0
    Stand for ditto, 10s. 6d.

3.  Taylor's Shallow-box or Eight-bar Hive, complete with
    cover  -       -       -       -       -       -  3  10  0
    Stand for ditto, 10s. 6d.

4.  Taylor's Amateur Bar Hive       -       -       -    3   5  0
    Stand for ditto, 10s. 6d.

5.  Neighbour's Improved Cottage Hive  -       -       -  1  15  0
    Stand for ditto, 10s. 6d.

6.  An Improved Cottage Hive       -       -       -    1   8  0

7.  The Ladies' Observatory or Crystal Hive.  Price complete  2   5  0

8.  The Cottager's Hive -       -       -       -       -  0  10  6

9 & 10.  Bee Feeders       -       -    - each 5/. and  0   4  0

11 & 12.  Fumigators       -       -       -  each 2/. and  0   2  6

| | | | | | | |
|---|---|---|---|---|---|---|
| 13. | Honey Cutters - | • | - | • | per pair | 0 5 0 |
| 14. | Taylor's Improved Cottage Hive | • | - | - | | 1 1 0 |
| | With Stand, £1. 10s. | | | | | |
| 15. | Fountain Bee Feeder | • | - | - | - | 0 6 0 |
| 18. | Taylor's Eight-bar Straw Hive, complete | | • | - | | 2 12 0 |
| | Stock Hive only, 15s. | | | | | |
| 19. | Huber's Book or Leaf Hive | - | - | - | | 2 5 0 |
| 20. | Taylor's Unicomb Observatory Hive | - | | - | - | 3 3 0 |
| 24. | Taylor's Glasses - | - | - | - | - 7/. | 0 4 6 |
| | Payne's ditto | - | - | - | - | 0 3 0 |
| 25. | Bell Glass | - | - | - | • | 0 4 0 |
| 26. | ,, | - | - | - | • | - 0 2 0 |
| 27. | ,, | • | - | - | - | 0 1 0 |
| 28. | ,, without knob and flat top to put on the table | | | | | |
| | inverted, price with lid | - | - | - | | - 0 4 6 |
| 29. | Zinc Cover - | - | - | • | each 7/6 and | 0 10 6 |
| 30. | Shallow Glasses (new shape) | - | | - each 5/6 and | | 0 3 6 |
| 31. | Bee Dress and Protector | - | | • | 5/., by post | 0 6 0 |
| 37. | Zinc Cover | - | - | - | - | 0 16 6 |
| 38. | Ornamental ditto | - | - | - | - | • 2 5 0 |
| 39. | Bee House to contain Two Hives - | | | - | - | 3 10 0 |
| 40. | ,, ,, Twelve Hives - | | | | £15. and | 19 10 0 |
| | Woodbury Unicomb Hive | - | - | • | - | |
| 42. | ,, Bar and Frame Hive, complete with outside | | | | | |
| | cover and super | - | | - | - | 3 3 0 |
| | Stand for ditto, 10s. 6d. | | | | | |
| 44. | Bottle Feeder | - | - | - | - | 0 2 6 |
| 45. | Woodbury Straw Bar and Frame Hive (Stock Hive) | | | | - | 1 1 0 |
| 46. | Engraved Pressing Roller | - | | - | - | 0 7 6 |
| 47. | Impressed Wax Sheets or Artificial Combs | | - | per dozen | | 0 6 0 |

www.ingramcontent.com/pod-product-compliance
Lightning Source LLC
Chambersburg PA
CBHW030905050726
47500CB00009B/1099